Charles Proteus Steinmetz

The Electrical Wizard of Schenectady

Robert W. Bly

Fresno, California

Charles Proteus Steinmetz

Published by Quill Driver Books
An imprint of Linden Publishing
2006 South Mary Street, Fresno, California 93721
(559) 233-6633 / (800) 345-4447
QuillDriverBooks.com

ISBN 978-1-61035-326-7

135798642

Printed in the United States of America
on acid-free paper.

Library of Congress Cataloging-in-Publication Data on file.

To William Oliver, my high school chemistry teacher, who encouraged in me a love of science that is almost boundless and has continued to intensify with each passing decade.

Contents

Acknowledgments

Thanks to my editor, Kent Sorsky, who recognized that Steinmetz, a giant in his field, is not as well known as the other electrical wizards of his day, and so deserves a new biography that I hope will make him more of a public name on a level with his contemporaries Westinghouse, Tesla, and Edison. Thanks also to T. C. Smith, Robert Lerose, and other friends who passed on Steinmetz material in their files to me. I enjoyed talking about Steinmetz with my friend David Justiss. And thanks to Carolyn Mazza, Belinda Robinson-Jones, Jodi Van Valkenburg, Nurit Mittlefehldt, and Fern Dickey for research and editorial assistance, and to Stephen Bly for clearing up some of the mathematics for me. Some of the photos of Steinmetz are from the collection of the Schenectady County Historical Society.

"No man really becomes a fool until he stops asking questions."
—Charles Proteus Steinmetz

Foreword

It is always a pleasure to watch your students succeed in their professional lives, and, for me, Bob Bly is no exception. Even though it has been more than 40 years since Bob was a student in my thermodynamics class at the University of Rochester, I remember him well. Over the years, while we were not in touch, I was well aware of his developing writing career (technical and otherwise), and I often read his blogs on the AIChE Discussion Central website. So I was pleasantly surprised when he asked me to add a few words as the foreword to this, his latest book. And given the focus of the book, I was more than happy to contribute.

In my opinion, Charles Proteus Steinmetz is one of the more remarkable scientific *and* engineering characters in the history of science and engineering—particularly here in America. But, also, he is one of the least publicized and, perhaps, least understood of those characters; which is why I am pleased Bob wrote this book.

Steinmetz was both a great scientist and an extraordinary engineer. He was, as Bob describes him, "A hardworking superachiever and a Renaissance dabbler." While Nikola Tesla has tended to get more public recognition (especially now with an electric car named after him!), Steinmetz really was the driving force behind the eventual universal adoption of the alternating current paradigm. By clearly describing and defining alternating current mathematically, he laid the foundation for virtually all that followed, even to our present day. His many contributions, including the identification and characterization of hysteresis, the creation of the magnetite arc lamp (the Steinmetz Street Lamp), and, especially, the development of the (now worldwide) power distribution grid, have impacted in a positive way countless millions of people globally.

But not as well known, Steinmetz was also a warm, generous, and sensitive human being who had to deal with significant physical adversity. One of the many aspects of Bob's book that I enjoyed was how he described Steinmetz as a man, as a human being. The relationship between Steinmetz and his beloved, adopted family is heartwarming and speaks to his true character, quite separate from his genius. The fact that he did not allow his physical challenges to negatively impact his life or his many interactions with others speaks directly to that character.

I enjoyed learning about Steinmetz's interesting home (and his private laboratory!) on Wendell Avenue in Schenectady, his unique rock garden, his lovely pool with fishes and an alligator, his Gila monster, and his love of flowers, plants, and photography. I expect that it is not well known that Steinmetz was an avowed socialist, although he seemed very comfortable working for a large, capitalistic company like General Electric.

Being an academic myself with nearly 50 years as a faculty member, a research scientist/engineer, and a university administrator, Steinmetz's love of teaching and research and his strong belief in a well-rounded education resonates well with me. In my opinion, these qualities help to differentiate him from many of his more well-known contemporaries, such as Ford and Edison. He was a prolific writer, authoring more than 200 publications and 13 books. I suspect this is another reason Bob became interested in Steinmetz.

Charles Proteus Steinmetz: The Electrical Wizard of Schenectady is well written, fascinating to read, and full of interesting details related not only to Steinmetz, but also to other scientists and engineers who, in turn, relate in one way or another to Steinmetz. I recommend this book to anyone with an interest in science and/or engineering or in the history of either, especially as it developed in America.

But I especially recommend it to those just beginning their careers (in any field, really) as an example of what can be accomplished when someone truly loves what they are doing and does not think of it as work. In other words, as Steinmetz did, find your passion and follow it, even though there will be many challenges to overcome along the way.

—Richard H. Heist, PhD
Senior Vice President, Emeritus,
Academic Affairs and Research
Embry-Riddle Aeronautical University

Introduction

The development of the electric power grid has until now been consistently portrayed as a two-man race between Thomas Edison, champion of direct current, and Nikola Tesla, a pioneer in alternating current. Sometimes George Westinghouse is brought into the fray, more as an industrial magnate (though he was also a talented inventor) who funded Tesla for a time.

But Charles Proteus Steinmetz, who once interviewed for a job with Edison's company but got turned away, barely gets a nod. Well known and well publicized in the late nineteenth and early twentieth centuries, when many people knew his name and revered him as an electrical genius—though Edison received far more press coverage[1]—Steinmetz is largely forgotten today, while new Edison and Tesla books continue to come out almost like clockwork.

Yet in many ways, Steinmetz was the most colorful and unusual of America's electrical pioneers. And he was easily the equal of Edison and Tesla, and arguably in some ways their superior, in his advancement of alternating current theory and development.

The Smithsonian has called Steinmetz "the wizard of Schenectady."[2] An article in an engineering journal proclaimed Steinmetz was "the father of electrical engineering."[3] The Electric Power Research Institute credited Steinmetz as "one of the great inventors and minds of the 1900s."[4] The website Electron Lab says, "Though Steinmetz has been largely overshadowed by Edison and Tesla, *his contributions to electrical theory are at least as significant*."[5]

Yet Steinmetz is barely mentioned in popular history books, while Edison and Tesla are known far and wide. This is in my opinion an egregious omission in the history of science and technology I hope this book in part will rectify.

Tesla was even featured as a character in the 2006 motion picture *The Prestige*, which fictionally portrays him as inventing a device that can replicate anything and make the duplicate materialize at a distance from the original—in a way, the first transporter.* And there is a fanbase of people who maintain a total fascination with all things Nikola Tesla.

But Steinmetz made contributions to science, or at least to the mathematics and machinery of electrical engineering, equal to or perhaps even greater than Westinghouse, Tesla, and Edison. All contributed original and essential work to create what is commonly known today as "the grid," the vast network of wire and cable that brings electricity from generators at power plants to homes, factories, and offices nationwide.

Well, Steinmetz deserves equal and maybe even top billing among these four giants of electricity: Steinmetz, Tesla, Edison, and Westinghouse. Yet he has been largely neglected for decades, with the exception of appearing as a character in Elizabeth Rosner's 2014 novel *Electric City* (Counterpoint Press).

The fascinating story of Steinmetz has two parts. The first is his brilliant work in the transmission of alternating current, which in many ways laid the foundation for the modern world as we know it. Steinmetz, Tesla, Edison, and Westinghouse did not invent electricity, the discovery of which actually predates Christ. But they made it possible for electricity to be delivered over great distances at reasonable cost to millions of businesses, schools, hospitals, and households; to light our world; and to power the machinery and technology on which it runs.

The second part of the Steinmetz story is the dramatic and moving tale of his personal life. Steinmetz had a brilliant mind, but his body was malformed: he was only 4 feet tall,[6] a hunchback, and had a defective gait, which made him walk with a noticeable limp. This, plus a head that was disproportionately large for his small frame, caused him to stand out physically from people with normal proportions, often drawing stares from children and impolite adult gawkers.

---

* The teleportation device in Clifford D. Simak's 1963 science fiction novel *Way Station* also teleports through duplication.

Steinmetz knew he had inherited these traits from his father, who in turn had inherited them from *his* father, Charles's grandfather. Determined not to make someone else live with the deformities he suffered, Steinmetz vowed never to marry and father children, so that a new Steinmetz generation would not inherit his genetic flaws.

But Steinmetz had great affection for children, and the second part of his story is how he found a family he could call his own, one that he eventually adopted, that gave him the companionship and love most human beings—including geniuses like Steinmetz—crave, though he never married.

It is only fitting that a man who provided so much comfort and pleasure to others in America, through the fruits of his genius and labors, was able in life to do, experience, and have the things he enjoyed most: his adopted family, who lived with him in his mansion in Schenectady, New York (his adopted grandchildren and their friends from the neighborhood flocked to the Steinmetz home and summer lake house); his friendships with brilliant colleagues such as Edison, Ford, and Einstein; his work; and his many hobbies and interests, including boating from his lakeside cabin and having a large greenhouse with a wide assortment of interesting fauna and flora, including lizards, turtles, and alligators.

Steinmetz was a mechanical and mathematical genius who was also a kind man, a good boss, a devoted family man, an educator, and generous with others. But he was serious about his work and wholly devoted to it; to Steinmetz, work was pure joy and an essential part of who he was. He was also a socialist, father, grandfather, friend, teacher, mentor, prolific author, and Renaissance thinker.

In this book, I want to introduce a whole new generation of readers, in particular those with an interest in science and technology, to a man who was changing the way we live through his mastery of electricity long before smartphones, personal computers, the internet, and digital networks existed. It was Steinmetz and his contemporaries whose works now provide the electricity that makes all of today's internet-of-things gadgets and smart technology—some of which are now digitally integrated with the modern electrical power grid—possible.

In addition to presenting the story of triumph over adversity resulting in a life well lived, I have one other concurrent goal in this book: to explain, in clear and simple language, the fundamentals in the developments of electricity leading up to and resulting in Steinmetz's groundbreaking work on the AC (alternating current) distribution grid—a ubiquitous power network that is nothing less than the foundation for how we live in the modern world.

I think it is by understanding not only the man but also the science behind his work that one can gain a total appreciation and enjoyment of the full story behind his life and times, as well as the brilliant mind, of Charles Proteus Steinmetz. Also, without understanding the electric grid, one cannot be said to possess a complete grasp of the way the industrialized and technological world works today, or of Steinmetz's monumental contributions to it.

I do have one favor to ask: If you think I have made an error or omission in the life of Steinmetz or the development of the modern electrical system, or a historical misstatement or technical error, please let me know so I can share it with readers of the next edition of this book. I can be reached at:

Robert W. Bly
31 Cheyenne Drive
Montville, NJ 07045
Phone: (973) 263-0562
Email: rwbly@bly.com
www.bly.com

1

Coming to America

Charles Proteus Steinmetz, son of Carl Heinrich and Caroline Neubert Steinmetz, was born in the old city of Breslau, Germany, on April 9, 1865. In 1889, at age 23, he immigrated to America to pursue a career as an engineer and scientist. But right from the beginning, his road was rocky and rough.

In Steinmetz's time, European immigrants crossed the ocean to reach the United States. Although immigration was lenient for much of the twentieth century, America was not always so welcoming to immigrants, especially those from different ethnic backgrounds or who were seen as mentally or physically inferior.

In 1889, Steinmetz came to America in the steerage of a French ocean liner, *La Champagne*, paying 75 francs for his one-way passage. After the boat unloaded the passengers, they were screened by immigration officers at Castle Garden, located at the southern end of Manhattan. This location continued to receive all immigrants until 1892, when Ellis Island opened and became the new entry point.[1, 2]

When he got off the boat, young Steinmetz immediately caught the attention of immigration inspectors, and in a negative way, because of his appearance: Steinmetz was a hunchback and a dwarf. Author Floyd Miller wrote: "[Steinmetz] was cruelly deformed. His spindly legs supported a thick torso with a hump that protruded high on his right shoulder; his overlarge head was covered with disheveled hair and his face with a bristly beard."[3] To make matters worse, a shipboard illness had caused Steinmetz's face to become swollen, adding to his unconventional appearance. As a result, Steinmetz was temporarily placed in a detention area and almost turned away. This increased scrutiny may have been in part due to a law passed in 1882, just a few years before Steinmetz's arrival, that banned entry of some immigrants into the United States because of poor health.[4]

The immigration official processing Steinmetz placed him in the holding cell with the intention of having him deported back to Germany, where he would be at risk of being sent to prison for being a socialist. But Steinmetz had a friend in Oscar Asmussen, a Dane whom he had first met in Switzerland and who had traveled with Steinmetz aboard the *La Champagne*. He, too, came here to become an American citizen and to make the United States his home. Asmussen, an experienced world traveler who passed through immigration without problem, approached the immigration officer and told him that Steinmetz was one of the most brilliant mathematicians in all of Europe and was here to consult with American businesses and professors about electrical theory. Asmussen also told the immigration official that Steinmetz was wealthy and that he (Asmussen) was carrying the scientist's money for him. Asmussen then removed a large roll of bills—actually his own money—from his pocket and gave some of the money to the immigration official as a bribe. Steinmetz's admission was approved, and together he and Asmussen entered the country.

Physical challenges for a mental giant

In addition to being just 4 feet tall and having a hump on his back (fig. 1-1), Steinmetz had an oversize head. Furthermore, he suffered from hip dysplasia, which caused him to walk with a shuffling gate. This combination of traits and conditions gave him an odd appearance that throughout his life drew stares, finger-pointing, and unkind comments. As mentioned in the introduction, Steinmetz decided against having biological children for fear of passing on the genetic defects he had inherited—hunchback, dwarfism, kyphosis, and hip dysplasia—from his father, a bookbinder of Jewish heritage, who in turn received them from *his* father.

Kyphosis is curvature of the spine: the spinal column is bent into the shape of the letter *C*, with the open part of the *C* pointing toward the front of the body. Everyone's spine is cured to some degree. But in a person with kyphosis, the curvature is severe, typically 40 degrees or more.

Severe curvature of the spine can cause a number of problems. We know with Steinmetz, for example, that the condition caused him to favor a bent-over posture, though with deliberate effort he could straighten up. Other common symptoms are nerve problems and chronic pain, which Steinmetz may have also suffered from, though we cannot be sure.

As noted earlier, Steinmetz's kyphosis was congenital, meaning he was born with the condition. However, extreme spinal curvature can also be caused by an accident, trauma, injury, or botched medical care or surgery. Osteoporosis, a weakening of bone often brought about with age and a declining ability to absorb calcium efficiently, can leave bones brittle and prone to fractures. It can also cause kyphosis. More common in women than men, osteoporosis is in fact now the most common cause of kyphosis in adults.

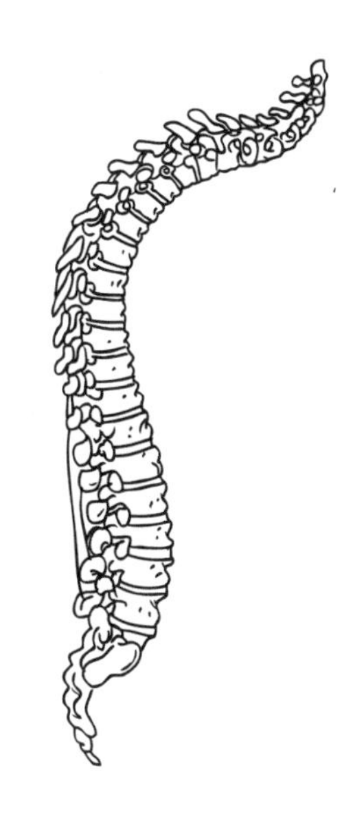

Fig. 1-1. Spinal curvature.

While hunchbacks are rare, they are not unheard of, both in real life and in fiction. Quasimodo, a character in Victor Hugo's novel *The Hunchback of Notre-Dame*, was a hunchback. Alexander Pope, the eighteenth-century poet, was a hunchback. More horribly, during World War II, Josef Mengele and other Nazi surgeons operated on Jewish concentration camp prisoners in their experiments, surgically turning physically normal men and women—including my own aunt Blanca, an extremely short woman not much taller than Steinmetz—into hunchbacks.

In Steinmetz's time, there were no practical treatments for severe spinal curvature. Later, some treatments were developed, including body casts and surgery.

Because his dwarfism was inherited, the cause was most likely a shortage of human growth hormone (HGH). Steinmetz also had a couple of the traits common to dwarfism, most notably a large head and the spinal curvature. Today, endocrinologists can examine a young child and accurately predict their adult height. If an endocrinologist determines that a male child will reach a full height of below 5 feet 5 inches in adulthood, he may be prescribed HGH injections; the parents give the shots at home,

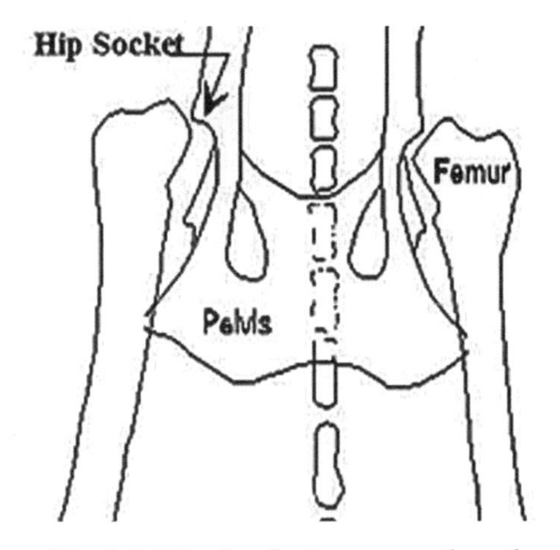

Fig. 1-2. Hip dysplasia occurs when the socket in the hip is too small to support the ball at the end of the femur.

and insurance often covers the cost. But HGH therapy did not exist in Steinmetz's time.

Added to Steinmetz's physical challenges was his hip dysplasia, a condition resulting from a deformity of the hip joint. Specifically, the hip socket in which the head or ball of the femur rests is too shallow (fig. 1-2), causing instability and excessive load on the rim of the socket.

His visible hip dysplasia symptom was a limp. Whether he also developed arthritis pain we do not know for sure, but most likely he did; many hip dysplasia patients suffer pain in the groin or side of the hip, which can cause discomfort when sitting, walking, or running. Because of this, when working, Steinmetz either stood or kneeled on a chair or stool, rather than sit on it.[5]

Despite his physical challenges, Steinmetz as an adult led a reasonably physically active life, enjoying hiking in the woods, canoeing on the lake at his summer cabin, bicycle riding, and roughhousing a bit with his adopted grandkids. Though he owned a car when he moved to Schenectady to work at General Electric, he most often rode his bicycle to work—he lived near the plant—and elsewhere around town.

His inherited deformity was the main reason why Steinmetz never married. What we are less sure of is whether his appearance and physical challenges also prevented him from marrying. Were women of the day repelled by a less-than-physically-perfect man? Of course, many people with physical challenges, from dwarfism to missing limbs, do get married and either have biological children or adopt. And it was in fact the adoption route that gave Steinmetz a family in his later years, as we shall see, despite not having a wife.

Child prodigy

Charles Proteus Steinmetz

Steinmetz showed his aptitude and even brilliance for scientific and technical thinking early. Many scientists in fact do their most important work early in their careers without ever matching those initial discoveries and achievements again later in life. This is especially true of theoreticians such as physicists and mathematicians. In contrast, other technical men and women, particularly those in engineering and other applied or practical scientific and technical pursuits, remain productive and creative throughout middle or even old age, though in most cases the creative spark dims, lessens, or goes out altogether eventually. Scientists from the nineteenth century who remained productive for many long decades include Steinmetz, Edison, and Tesla, to name just a few.

Some scientists and engineers are mainly hands-on rather than theoreticians. Edison is an example. When asked by a reporter his opinion of Einstein's theory of relativity, Edison replied curtly that he had no opinion because he did not understand it (though later in life he seemed to take more of an interest in Einstein's work, particularly the conservation of energy).

Other scientists and engineers are theoreticians who are more comfortable in front of the chalkboard or computer writing equations than at the workbench or holding a soldering iron. Theoretical physicist Einstein was in that category. All his experiments were "thought experiments," meaning he carried them out with pure thought and imagination, with no laboratory work or physical measurements or observations performed in the field.

Steinmetz in this regard was successful in both worlds, equally at home with complex mathematics and working in his laboratory tinkering, experimenting, and observing. Steinmetz said, "All we know of the word is derived from the perception of our senses. They are the only real facts; all things else are conclusions from them."[6] But he also attributed his great success in science and engineering to his thorough study of and strong abilities in mathematics.

In science and engineering, Germany is best known for its excellence in rocketry, chemistry, and physics, and Steinmetz was brilliant in mathematics and electrical engineering. While still in high school, which was called "gymnasium" in Germany at the time, he wrote technical papers for his science classes that were quite advanced for someone his age, which his father the bookbinder bound between covers for posterity.

A child prodigy, Steinmetz entered gymnasium when he was just eight years old. There he enthusiastically threw himself into learning Latin, French, Greek, Hebrew, Polish, mathematics, logic, philosophy, and the classics. He had a prodigious memory, and after studying Horace and Homer he could, for the rest of his life, recite long passages from these ancient poets. As an adult, he believed everyone should have a well-rounded education in the liberal arts, and wrote an article for an engineering journal saying so. For all of his adult life, Steinmetz railed against universities where science, engineering, and mathematics majors took only technical courses, and he advocated that engineers should receive a well-rounded education with more emphasis on languages, literature, history, and other liberal arts subjects.

Steinmetz graduated from gymnasium in 1883[7] with such high marks that he was not required to take oral examinations. That fall he enrolled in the University of Breslau, one of the few schools in Germany that included the study of electricity in its physics curriculum. At the time, no German schools offered electrical engineering as a major or even had courses in the discipline. Today, electrical engineering has flourished as a course of study and a profession. Its trade association, the Institute of Electrical and Electronics Engineers—abbreviated as IEEE and known as "eye-triple-E"—has over 400,000 members in 160 countries, making it the largest professional society of engineers in the world.

In Germany, which so favored scientists and engineers, why didn't Steinmetz remain in his homeland to practice his science and engineering? It was largely because of the Jewish heritage on his father's side: Though the full force of anti-Semitism did not strike German Jews until Hitler rose to power, reaching its peak with the mass extermination of 6 million Jews during World War II, many historians say that Germany had a history of discrimination against Jewish people long before that. This anti-Semitism had its beginning in the Middle Ages, when many Europeans resented the Jews for their success in business. The prejudice intensified in the 1800s as more and more Jewish people entered the mainstream of European society.

In addition to being a Jew, young Steinmetz was a socialist—a political and ideological stance that gave offense to the German authorities. The Manifesto of the Socialist League, published in 1885, gives this overview of the socialist principles that Steinmetz sympathized with:

> As the civilised world is at present constituted, there are two classes of Society—the one possessing wealth and the instruments of its production, the other producing wealth by means of those instruments but only by the leave and for the use of the possessing classes.
>
> These two classes are necessarily in antagonism to one another. The possessing class, or non-producers, can only live as a class on the unpaid labour of the producers—the more unpaid labour they can wring out of them, the richer they will be; therefore the producing class—the workers—are driven to strive to better themselves at the expense of the possessing class, and the conflict between the two is ceaseless. Sometimes it takes the form of open rebellion, sometimes of strikes, sometimes of mere widespread mendicancy and crime; but it is always going on in one form or other, though it may not always be obvious to the thoughtless looker-on.
>
> The profit-grinding system is maintained by competition . . . not only between the conflicting classes . . . there is always a war among the workers for bare subsistence, and among their masters, the employees . . . for the share of the profit wrung out of the workers.[8]

Steinmetz became a member of the Social Democratic Party, which the German government viewed as politically subversive. When the editor of the party's paper was arrested and imprisoned, Steinmetz took over, though he was smart enough not to write political articles under his real name. For several months, he edited the paper and also two other periodicals, *Popular Science Leaflets* and *Popular Science Fortnightly*. As a result, his membership in the socialist party was not exactly a well-kept secret, and so the government placed Steinmetz under surveillance, eventually ordering his arrest.[9]

In the 1880s, Germany's Chancellor Otto von Bismarck opposed socialism. He controlled the military and the national bureaucracy, giving Steinmetz another reason to want to leave, despite Germany's rapid industrialization and encouragement of engineering and scientific innovation. Steinmetz would later play a vital role in industrialization, but as

Otto von Bismarck, chancellor of the German Empire from 1871 to 1890.

an American citizen, not a German.

Although Charles Steinmetz remained a socialist for life, in the United States he enjoyed a long and fruitful career as an employee of a large corporation, the type of entity that socialists frown on. His reasoning was that General Electric was efficient and well organized, paid its workers at all levels fairly (for the most part), and made possible great advances in science and technology that benefited humankind.

Those were frightening times for Steinmetz, fraught with peril. He had several close brushes with the police and was continually under suspicion, and there were efforts to exclude Steinmetz from the university. The situation became so unbearable and tenuous that Steinmetz left Germany before he could complete his PhD studies and by doing so likely avoided arrest. He fled across the Austrian border, traveling first to Switzerland, where he lived and studied for a time, before going on to the United States. It was in the United States where he finally settled and flourished, making major contributions to America through his technical work, and ultimately sharing it with the world in the electrical power system that runs our infrastructure today.

During his time in Switzerland, Steinmetz had little or no money and no immediate, visible means of support. But throughout his life, and starting early, in addition to being a productive inventor, Steinmetz was a prolific author. He had brought with him to Switzerland the manuscript of a book he had written, not on electrical engineering but on astronomy. A Swiss publisher paid him 37 francs a month for it, which gave him enough money to get by. He wrote more articles on astronomy for a Swiss newspaper, which paid him an additional 15 to 20 francs a month. His productivity for a time in this field was remarkable because, although he had taken at least one astronomy course at university, he was not an astronomy major nor did he do significant scientific work in the field.

And so initially after fleeing Germany, Steinmetz earned income as a writer of popular science. This may explain in part why he was always patient with newspaper reporters and magazine journalists who came to interview Steinmetz for articles about his work. Most had no science background, but Steinmetz was able to explain his work in clear language that even a nontechnical person could easily understand.

In 1888, Steinmetz entered the polytechnic school in Zurich. He studied mechanical engineering, steam engines, bridge construction, and turbines; knowledge of the latter came in useful later when turbines served as a key component of the electrical grid system that Steinmetz helped design as chief engineer for the General Electric Company.

An American engineer in Yonkers

Though there was and still is much anti-Semitism both in the United States and overseas, one advantage Steinmetz had moving in the late nineteenth century was that making the transition from foreigner to U.S. citizen was quicker, simpler, and somewhat less restrictive and less bureaucratic than it is today. So he was able to become a citizen in short order.

His German birth certificate listed his full name as Karl August Rudolf Steinmetz. When Steinmetz applied for American citizenship, he changed his German first name Karl to the American Charles, adding the middle name Proteus, after the Greek sea-god who was the son of Poseidon.

In the United States Steinmetz was hired as a draftsman, his first job, for $12 a week. Of course, this was more than a century before the invention of computers and CAD (computer-aided design) software, so engineering drawings were all rendered by hand. Engineering students were typically required to take drafting class, and Steinmetz was no exception. I took drafting in high school, and unlike today, when the computer makes corrections easy, if you made a mistake by hand, you had to carefully erase. And if the mistake was too big, you drew the whole thing over again. But from the blueprints, device diagrams, circuit diagrams, and patent application drawings Steinmetz produced, we clearly see his proficiency in drafting.

Steinmetz's employer was Eickemeyer and Osterheld (E&O) in Yonkers, New York, an industrial firm owned by Rudolf Eickemeyer, who was an electrician by trade. E&O was best known for making high-quality transformers, which are devices used to either increase or decrease the voltage of electric current. E&O was a small but active research and development (R&D) company, developing public transportation and motors, among other products. Between 1856 and 1880, Eickemeyer was awarded 67 patents, 50 of which were, oddly enough, for hat-making machinery and ancillary equipment. When Alexander Graham Bell invented the telephone in 1876, Eickemeyer began developing and patenting improvements on Bell's telephone, as did Thomas Edison.[10] What helped Steinmetz get the E&O job was a recommendation written to Eickemeyer from a Mr. Uppenhorn, who was the editor of a German technical publication for which Steinmetz had written several articles on electrical topics.

One day, Mr. Eickemeyer accidentally spilled ink on his hands; it was aniline ink, which is very quick-drying, and so it would not come off with soap and water. His new employee Steinmetz quickly mixed a simple, nonharmful chemical solution, which he then used to safely and completely remove the ink from his boss's skin, with no burning sensation or pain.

Whenever you go above and beyond the call of duty at work, or solve a problem no one else seems able to, you attract positive attention, and your star at the company can rise. Steinmetz's quick thinking and practical chemical knowledge impressed Eickemeyer, who realized he had a potential gem in his employ. Eickemeyer made a smart management decision, promoting Steinmetz to head of research, which put him in charge of the company's experimental work. Meanwhile, he allowed Steinmetz to continue his technical studies at night as well as write scientific papers.

Like Steinmetz, Eickemeyer had similarly been persecuted by the German government. And like Steinmetz, only many years earlier in 1848, he had immigrated to America, where he flourished as a successful business owner and entrepreneur. As would be the case years later when Steinmetz was the boss of engineering at GE and befriended his young lab assistant Joe Hayden, the more senior Eickemeyer became both a valued colleague and a friend, and Steinmetz was soon a regular dinner guest at the wealthy industrialist's mansion.

Toward the close of the nineteenth century, Eickemeyer's company was actively developing electrical motors for industrial equipment. One of these industrial applications was a motor to propel an electric car. Although today we think of electric cars as a recent invention, in fact they were first built in the 1800s. The electric cars Eickemeyer's firm designed were for mass transit: streetcars carrying passengers, not private cars to be owned by individual drivers. Steinmetz was told to make drawings for various parts of Eickemeyer's electric streetcar, which was in essence a trolley car. These included an overhead connector and switch.

Other drafting assignments given to Steinmetz by Eickemeyer included electric water pumps and a magnetic-ore separator; the latter was a direct competitor with Thomas Edison's ore separator, which was used at Edison's milling plant in Ogden, New Jersey. Many of Steinmetz's diagrams were drawings used in patent applications. After about half a year of working mainly as a draftsman, Steinmetz took on additional engineering design work for the new streetcar system.

By 1890, E&O had a working design for a commercial streetcar that featured a gearless motor. They made the drive wheel as large as the front wheel, reduced the speed of the armature shaft, and boosted the terminal power to 500 volts DC (direct current). That same year, the first eight cars built ran on five and a half miles of track owned by the Lynchburg Street Railway Company. The cars ran well even on steep grades. Steinmetz then designed a switch and copper brushes for the motor and did the calculations for sizing the power cables.[11]

Next, Eickemeyer told Steinmetz to design a motor for a new type of machine called a vertical trolley, a small car that conveyed passengers vertically within a building rather than horizontally. The manufacturer called it an elevator, after the equipment's primary function of elevating what it carried, and the maker soon named his firm Otis Elevator Company.

The motor Eickemeyer and Steinmetz designed for their customer Otis was of a "shunt-wound" (fig. 1-5) direct current configuration, the first of its type ever produced. Electrical current can be generated by passing an armature wound with copper coils through a magnetic field.

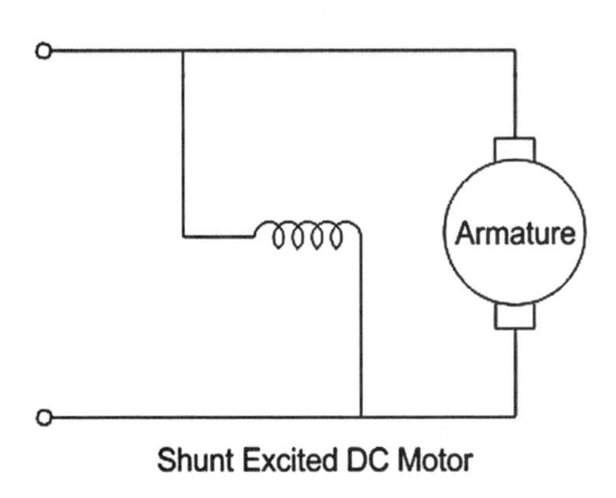

Fig. 1-5. Shunt motor.

When the windings and armature are connected in a parallel combination, the design is called, in electrical terminology, a shunt.

After helping to design the world's first elevator motor, Steinmetz was tasked with the design of both electric water pumps and dynamos. A dynamo is a generator that produces direct current. In a dynamo, a metal ring or cylindrical disk is rotated on an armature. The ring or cylinder is in contact with metal brushes that collect the current generated as the armature spins.

Eickemeyer's company was so successful, thanks in large part to Steinmetz's superior engineering design work, that the then recently formed General Electric Company, headquartered in Schenectady, New York, bought out Eickemeyer in the early 1890s. Along with the various assets of the firm, including the patents on all of Steinmetz's designs, the deal called for Steinmetz himself to continue his work as a General Electric (GE) employee, which Steinmetz agreed to. No surprise there; he clearly loved the work.

"Steinmetz had the creed of a searcher," wrote John Hammond in his book *Men and Volts: The Story of General Electric*. "He was never tired of laboratory work. That was where happiness lay for Steinmetz. He lost himself for long hours, for half a night, in his private laboratory in Schenectady, and what he did there he never alluded to as 'work.'"[12]

In point of fact, the main reason GE bought E&O was not for the patents or for the sales; it bought out Eickemeyer to acquire Steinmetz. GE had previously tried to hire Steinmetz away from Eickemeyer, but Steinmetz was loyal to the man who gave him his first job in America when he needed it, and he refused GE's offer, despite the considerably higher salary. The only way GE could get Steinmetz was to buy out the company he was working for—and GE did just that. It turned out to be a corporate acquisition that generated a return on investment almost beyond measure.

Ironically, General Electric had been formed from Edison Machine Works, a company started by Thomas Edison, who as we shall later learn became one of Steinmetz's fiercest rivals in the development of America's electrical power distribution system. In fact, before going to work for Eickemeyer, Steinmetz had interviewed for an engineering position with Edison's firm but was not hired.

When Edison was interviewing new employees, the story goes, he handed them an elaborate glass vessel, with all kinds of shapes and extensions, filled with water and told them to figure out how much water was in it. Some candidates spent a half hour or more measuring and calculating volume, but because of the complex shape they were invariably wrong. Successful candidates, on the other hand, used a simpler approach. These engineers simply took a graduated cylinder from Edison's lab bench, emptied the water from the odd-shaped vessel into it, and instantly told Edison the exact volume. These were the men Edison hired. There is no record of whether Steinmetz was given this test.

At GE, Steinmetz and his ideas flourished. He was the company's star employee in the field of electrical engineering research and development. His ideas and designs contributed greatly to GE's growth and success, and Steinmetz was rewarded with promotions until he eventually became the chief engineer in charge of the R&D department.

He was continually offered raises, though accounts vary on how much he actually got paid. Some say that at one point the company just gave him a blank check every month on which Steinmetz could write whatever sum of money he wished. However, this is extremely unlikely, as large corporations operate under strict financial control and would never give one employee, especially a technical man and not a financial officer, such control over disbursements. One time when offered a raise, Steinmetz told management to give the money instead to increasing the salaries of his lab assistants. By some accounts, he was at one point offered a salary of $100,000—a fortune in those days—but other histories say this is unlikely.

After working for both a small business, E&O, and a giant corporation, GE, Steinmetz wrote that in his view, the large industrial corporation was an institution superior to small business in productivity, innovation, wealth-building, and treatment of the labor force:

> The industrial corporation of to-day is organized for effective constructive work; it has developed the characteristics necessary for economic efficiency—continuity of organization and at the same time flexibility to adapt itself in a high degree to the requirements of industrial production, and to the personality of its members; it has within itself the responsibility of the individual toward the whole, and encourages initiative and individualistic development as important factors of industrial progress, and especially it has solved the problem of filling the offices with competent and qualified men. Neither the political Government nor any other organization has these characteristics, and it therefore appears the natural and most logical step that the executive and administrative Government of our nation in the co-operative era which we are now entering should evolve from the industrial corporation.

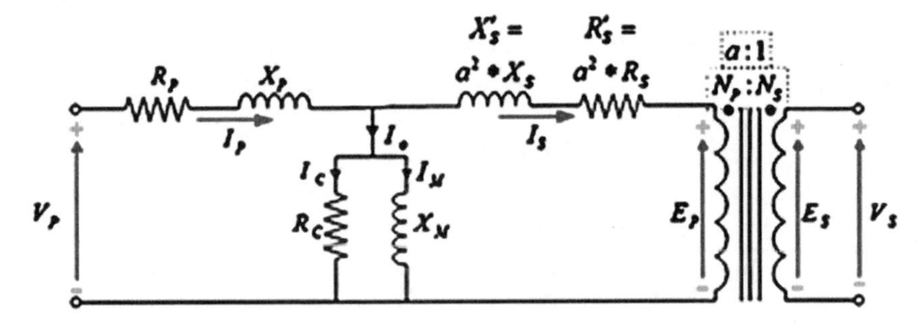

Fig. 1-6. An electrical circuit designed by Steinmetz.

> Such organization is commensal—that is, every member of it gives and receives, and the maintenance and advance of the organization thus is to everybody's interest. It thus should form a stable and permanent form of society, permanent at least as long as the foundation of our civilization endures, as stable as was the classic age or the feudal age of human society, and not self-destructive by its own success, as was the individualistic age.[13]

Perhaps Steinmetz's greatest achievement at GE was to rigorously work out the mathematics governing the characteristics and transmission of alternating current and AC circuitry. He achieved the twin goals of reaching a deeper understanding of the science behind electrical current and discovering how to build better AC circuits to transmit electricity over power lines farther and more efficiently (fig. 1-6). By "efficient" power transmission, we mean transmitting the current great distances while experiencing the minimal loss of power over the cables.

Real power from imaginary numbers

In his mathematical analysis of AC circuitry, Steinmetz used complex numbers involving the "imaginary number," which is the square root of negative one. He referred to the square root of negative one as "The General Number" when he first presented his calculations of alternating current at a meeting of the International Electrical Congress in Chicago, Illinois, in 1893. In particular, Steinmetz used imaginary numbers to describe impedance, which is a resistance that prevents or impedes alternating current from flowing easily through an AC circuit. (In direct current, this barrier to electron flow is simply called the resistance of the circuit.)

Now, a number that is the square root of negative one may sound wrong-headed to you. After all, the square root of a number is the number that, if multiplied by itself or squared, yields that number. For instance, the square root of four is two, because two times two equals four.

The square root of negative one is called an imaginary number, because negative numbers have no real square root. The reason is that a negative number multiplied by itself or any other negative number, at least in "real" mathematics, always yields a positive. There is no number that, multiplied by itself, produces a negative number. Yet mathematicians routinely use imaginary numbers in complex calculations to solve equations that can be solved only with the use of those imaginary numbers. One science writer noted that Steinmetz "essentially made electricity out of the square root of negative one."[14]

The idea of creating numbers that work in equations but can be said not to exist in real life applies to numbers much simpler than imaginary numbers. Take familiar, everyday negative numbers, for instance. Negative five seems sensible enough. But can you have negative five people in a room? One can argue that there is no such thing. Yet no one seems to have a problem with ordinary negative numbers.

Steinmetz, ever the author, set down his ideas about alternating current in a series of books, though these are aimed at engineers and other technical readers, not the general public. He detailed the fundamentals of electrical engineering, including conduction, magnetism, wave shape, resistance, and stability, in his book *Theory and Calculation of Alternating Current Phenomena*, first published in 1897. On its first page, he notes:

> When electric power flows through a circuit, we find phenomena taking place outside of the conductor which directs the flow of power, and also inside. The phenomena outside of the conductor are called the electric field. Inside of the conductor, we find a conversion of energy into heat; that is, the electric power is consumed in the conductor by what may be considered as a kind of resistance of the conductor to the flow of electric power, representing the power consumption of the conductor.
>
> In other words, electrons flowing through a copper wire generate a magnetic, or more properly an electromagnetic, field, which exists outside the wire or cable and is strongest in its immediate vicinity. Today we have established that electromagnetic fields (EMFs) can directly affect living cells, in particular their growth. Sometimes the growth is affected in a beneficial way, such as accelerating the healing of broken bones. Other times, the growth can be harmful, as in cancer, which is why so many people today avoid buying homes or renting apartments close to overhead power lines.

Electron flow within the copper wire, in addition to generating electromagnetism, also produces heat in the wire or cable through which it travels. This is how we get light and heat from light bulbs, and why the wires in a toaster glow orange and generate heat to toast our bread. Electrical space heaters in which the coils turn orange work in much the same way as the toaster.

Metals are good conductors of electric current. Why? Because metals have a high number of movable, or free, electrons. These wandering electrons can pass easily through the metal,[15] even as the nuclei of the metal atoms are held firmly in place by strong atomic bonds.[16] Pure metals are better conductors than alloys; and both types of metal are better conductors than electrolytes, such as salt water [fig. 1-7].

As far as pure metals go, silver is the best conductor. However, silver, an investment-grade metal, is far too expensive to use in wiring. Copper, though not quite as conductive as silver, is nevertheless an excellent conductor, while being much more affordable. Copper is also strong, reliable, bendable, and easy to solder and form into wires.[17]

Steinmetz's books are considered the basis of a complete technical education in the fundamentals of electrical engineering in general and alternating current in particular. In addition to his technical writings, Steinmetz held more than 200 patents for inventions in every phase of electrical engineering.

In America, Steinmetz's reputation as a leading expert in electrical phenomena solidified and grew quickly. At General Electric, he was known as a brilliant innovator, a happy workaholic, and a bit of an eccentric. For instance, Steinmetz loved cigars and smoked them throughout the day. Upper management at GE decided one day to no longer allow smoking within the plant. When told he could not light up his cigars at work, Steinmetz replied: "No smoking, no Steinmetz." Willing to do anything to

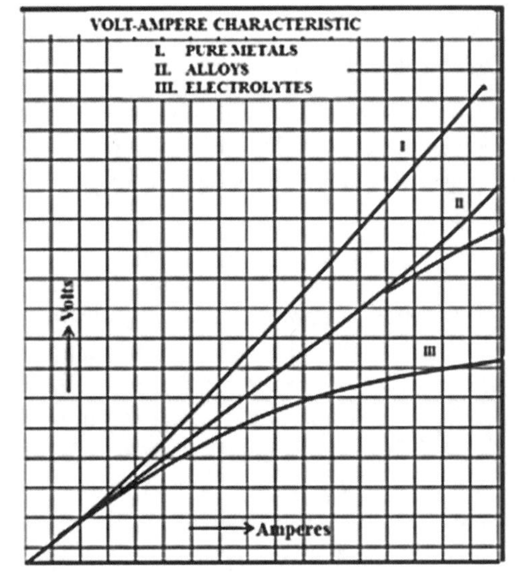

Fig. 1-7. Conductivity of metals is superior to electrolytes.

please its most valuable employee, who clearly insisted on having his way on the smoking issue, GE management recanted and allowed Steinmetz to smoke all the cigars at work he wanted. But smoking on the job remained absolutely forbidden for everyone else. Steinmetz bought special custom cigars made with a blend of tobacco that he said was a milder smoke.

Steinmetz's work was of paramount importance to him. He continually thought about and solved problems in the mathematics of electricity and electrical distribution while at his office at GE, in a private lab in his spacious home on Wendell Avenue near the GE plant, and also while sitting in his canoe as he drifted across the placid surface of Lake Mohawk, where he had a summer home.

Steinmetz remained a staunch socialist throughout his life. Though he worked for a large corporation in a capitalist society, he did not embrace capitalism philosophically. He once ran for state office on the socialist ticket, though he did not win. The socialist worldview does not place a great value on the accumulation of wealth, and neither did Steinmetz. At GE, or so one story goes, he could have demanded and would have been given a much larger salary than the one he actually took. Most likely he did in fact get a large salary, but the reason he did not earn even more was precisely that he was so content with his position at GE that he never asked for more money.

Money was not a primary motivator for Steinmetz. What he valued most from GE was that the company gave him the freedom to do his work in peace and in the way he saw fit. And management, knowing how happy he was there, figured they didn't have to offer a bigger salary to keep him.

His goal was not the accumulation of wealth. Nor did he seek fame or accolades. He accepted positions at both a university and a professional society, not to feed his ego but to help advance the state of the art in his chosen field, electrical engineering, and teach others what he knew, a compulsion that also drove him to be a prolific author of books and papers.

Rather, his objectives in life were fourfold: First, to do his science. Steinmetz had a brilliant mind, and like many highly intelligent people, he seemed unable to stop thinking about his work, even when at his country cabin for rest and relaxation. He frequently worked on engineering and math problems for 12 hours a day or longer.

Second, and related to the first, was his desire to pursue and enjoy his many other interests. Some, such as cycling, were related to his enjoyment of nature and the outdoors. But many others, such as botany and exotic animal keeping, engaged his intellectual curiosity. He wrote books on topics ranging from mathematics and astronomy to politics and history. It seemed there were no subjects in which he could not get interested.

Third, he sought freedom. In Germany, as a Jew and a socialist, he was always under the watchful eye of a hostile government that often considered him a subversive and an undesirable. In America, he had escaped the tyranny of Bismarck. Here, he enjoyed the freedom to come and go as he pleased, without surveillance or fear of the police, and lived a life he wanted to live. He indulged in politics; dedicated himself to his work; built a large, comfortable home; and located the house right next to the GE plant to avoid a long commute. He also equipped the house with a laboratory that rivaled his GE workshop so he could go to his workbench and experiment at any time of the day or night, whenever the mood struck him—and it struck almost all the time.

Fourth, he wanted relationships and companionship. He especially enjoyed meeting and befriending other men in his and related fields of science and engineering, including Einstein and Edison. But more than friends, he wanted a family. As mentioned previously, because of his genetics he decided early not to have biological children. And perhaps because of his appearance, he felt no woman would have him. But despite these drawbacks, as we shall see later in the book, he achieved at least the former, having a household filled with an adopted son and his son's children. He became a father and grandfather, which gave him the love and companionship he so fervently desired.

2

The Evolution of the Electric World

Our amazing power grid

If you look around you, you can't help but notice how electricity plays an essential role in almost every facet of modern life. For instance, I am writing this book on a PC, which requires electricity to operate. Disconnect the computer from its power source, and the screen will go dark after the uninterruptible power supply (UPS) runs out of voltage—and what I am writing might be lost if I have neglected to back it up.*

In my home office, even though I begin work early when it is still dark outside, I can see thanks to the light bulbs in the ceiling, whose filaments glow because of the electric current running through them.† The office is cooled in summer by a central air-conditioning system that runs on electricity and heated in the winter by a furnace where the fuel, natural gas, is ignited by an electric spark. As I write, I am listening to music on a CD player that is plugged into a wall outlet. Similarly, my printer, photocopier, and fax machine would be dead hunks of plastic and metal without household current, brought from the utility power plant to my home by the electric distribution grid that Charles Proteus Steinmetz was instrumental in developing.

The rest of the house, from ceiling, floor, lamp, and outside lights to the garage door opener, dishwasher, microwave oven, coffee pot, and TV, also depends on electricity. The refrigerator and freezer need current to operate.

---

* Actually I now have a cloud-based automatic backup, though I am not 100 percent confident it works in that I cannot see it.

† One desk lamp has an oversize glass bulb, not frosted but completely clear, displaying the large bright filament inside when the lamp is switched on.

In case of a blackout, your food will soon spoil. Without household electric current from the grid, the only way to charge mobile phones is with a charger plugged into a car with the motor on.

So, what exactly is "electric current"? Any substance showing an attractive force is said to be electrified or to have an electric charge. Electricity is simply an electrical charge moving from one point to another. Our dependence on electrical current and on the grid system for delivering electricity is dramatically demonstrated when a storm, downed or cut power line, or other condition results in a power outage in your house, neighborhood, or town. We call the electric utility, outraged that this has happened and demanding restoration of power instantly. In the United States, we have become dependent on electricity—even spoiled by its almost constant availability—and we are quick to anger when we don't get our power back lickety-split.

People in many other nations are not as fortunate as Americans when it comes to continuous access to reliable electricity—or any electricity at all. In India, 400 million people do not have electricity because the Indian power grid does not extend to their homes.[1] In Haiti, only 34 percent of the population has reliable power.[2] In 36 countries across Africa, 600 million residents have no electricity.[3] And the power supply for those who do have electricity is unreliable. In Nigeria, most of the households are connected to the power grid, but in four out of five homes the electricity works less than half of the time.[4]

Worldwide, 1.3 billion people lack access to electricity.[5] Yet if the typical American loses power for more than a few minutes, he or she carps about it indignantly until the problem is resolved.

But now, in America, power outages are likely to occur with increasing frequency. From 2000 to 2004, there was an average of 44 reported grid outages annually. From 2010 to 2013, there were on average 200 grid outages per year.[6] Many scientists believe global warming has disturbed natural weather patterns, resulting in devastating super storms such as Hurricane Katrina in Louisiana, Hurricane Sandy in New Jersey, Tropical Storm Harvey in Texas, and Hurricane Irma in Florida. The high winds rip branches loose and sometimes even knock over trees, which fall into the power lines connected to your home, dangerously dislodging them. In 2017, approximately 140,000 people[7] were left without power in the San Fernando Valley, California, when the local power plant had a catastrophic failure.[8]

Massive solar flares can also disrupt electrical power. In 1859, a solar storm shut down the entire telegraph network of Europe and North America. In July 2012, NASA reported that the energy output from the most powerful solar event in 150 years narrowly missed striking the earth. Had it hit, there would have been chaos equivalent to 20 Hurricane Katrinas.[9]

So dependent are Americans for nonstop, 24-hours-a-day, 7-days-a-week, 52-weeks-a-year electric power that a power failure causes significant distress, unhappiness, even panic, and, in major cities, looting and other crimes. The Federal Energy Regulatory Commission (FERC) defines the "resilience" of a power plant or grid as "the ability to withstand and reduce the magnitude and duration of disruptive events . . . and rapidly recover from such events."[10]

"Our society simply was not built to function without electricity," says science writer Jared Brewer. "Imagine living two months without lights, without refrigeration, without a cell phone."[11] In René Barjavel's 1943 science fiction novel *Ashes, Ashes*, when all electricity is suddenly cut off in a futuristic high-tech world, society collapses into anarchy.[12] I have a friend who works for a major power utility in New York, and during storms, he and his crew put in marathon hours, often outdoors in thunderstorms and blizzards, to make sure the power stays on, and to restore it as rapidly as possible when it goes out.

Emergency power to the rescue

To meet the ever-growing demand for uninterrupted electric current and to keep the lights on, a new home generator industry has mushroomed as people prepare for the next Henry or Irma.

Before the widespread commercial availability of both portable and permanent backup generators, when the power went out, homeowners were advised always to have flashlights. But they often forgot to change the batteries, and so when the lights went out, they clicked the on switch for the flashlight—and nothing happened. And in the dark, spare batteries, if any, were hard to find. Many people keep candles for light during power failures because candles work without batteries, producing light by burning the wax and wick, both of which are consumed by the flame. But candles are not as safe as flashlights and cause nearly 2,000 house fires each year in America.[13]

Traditional home generators, which have been around much longer than whole-house units, are small, portable gasoline-powered units. They generate limited current and can only power select devices in the house;

the generator is typically hooked up to the freezer, refrigerator, a radio for news of the storm's progress, and some table lamps for lighting.

But portable generators are problematic. For one thing, they hold only a small amount of gasoline that they consume rapidly, so to get more gas you have to drive out in the storm to find a local gas station that still has fuel and operating pumps—usually in another town that still has power. But during Hurricane Sandy, in my part of New Jersey, the lines to get gas at those stations were often two hours long, and during Henry and Irma, the roads were so flooded that driving to pick up gasoline became nearly impossible.

A second problem with gasoline-powered generators is that they emit carbon monoxide, which can poison and suffocate the unwary user who operates his generator in a closed space. The best place for your portable generator is near the entrance to your garage, with the door and any windows open. When you do this, you need a long extension cord to power the devices and appliances inside your house. And when you stand near the open door to operate the generator in stormy weather, you are going to get wet.

Today Generac, Cummins, Kohler, and General Electric—the company that employed Charles Steinmetz for most of his career—are selling a better backup power source known as the whole house generator. These are larger units permanently installed outside the home. They can be powered by natural gas if your house is supplied with natural gas by your local utility. If not, propane also works. But natural gas is better: propane tanks eventually run out, and additional propane may be difficult to buy when the power is out and everyone needs a tank.

Whole house generators afford a welcome degree of protection against power outages, but they are not cheap. For our 3,300-square-foot home, our unit cost almost $13,000 when installed in 2015. Sales of generators in the United States are forecast to reach $24.5 billion by 2021.[14] That's how reliant we are on 24/7 electricity and how reluctant we are to part with it.

The big emerging technology in emergency power today is now the backup electric grid. For instance, the town of Fujisawa, Japan, has a backup renewable energy system that can provide off-grid power for up to three days. As I write this book, a similar backup grid is being built on a 400-acre site near Denver International Airport. In the Denver project, a solar grid with lithium-ion storage batteries will provide backup power.[15]

Who discovered electricity?

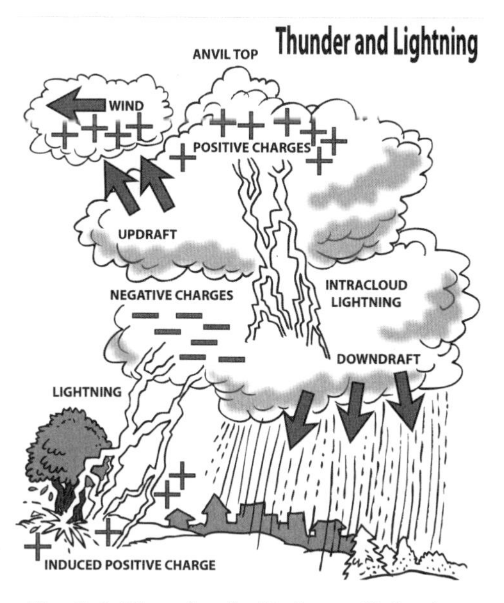

Fig. 2-1. How clouds discharge lightning.

Contrary to what many of us were taught in grade school, Benjamin Franklin—scientist, statesman, entrepreneur, and printer—did not discover electricity. The story of Franklin flying a kite by tying a metal key to it and standing in a thunderstorm is true. Franklin's famous kite experiment, conducted in 1752, showed lightning and electricity were in fact the same thing.

Violent movement of ice crystals and raindrops in storm clouds causes electric charges to build up, and the clouds become charged with positive particles, protons, and negative particles, electrons (fig. 2-1). Cloud-to-cloud, or "intercloud," lightning can cause the positive and negative charges to become unbalanced in the two clouds exchanging bolts of electricity.

When an excess of electrons, the negatively charged particles, builds up in the bottom portion of one of the clouds, that cloud can then discharge a bolt of electrons that strikes the ground or something on the ground. Charles Steinmetz studied lightning in-depth and built machines capable of generating artificial lightning from condensers, using a principle similar to clouds.

In Franklin's famous experiment, a metal key was attached to the bottom of the damp kite string. A silk ribbon tied to the string insulated Franklin's hand. When a lightning storm struck, electricity traveled down the wet string to the metal key. The electrical charge of the lightning was stored in a Leyden jar, a device consisting of a brass rod inside an insulating glass container with a wood or cork lid, and coated inside and outside with metal foil (fig. 2-2).

More than likely, Franklin's kite was not hit directly by a lightning bolt, which in all probability would have seriously injured or even killed him.

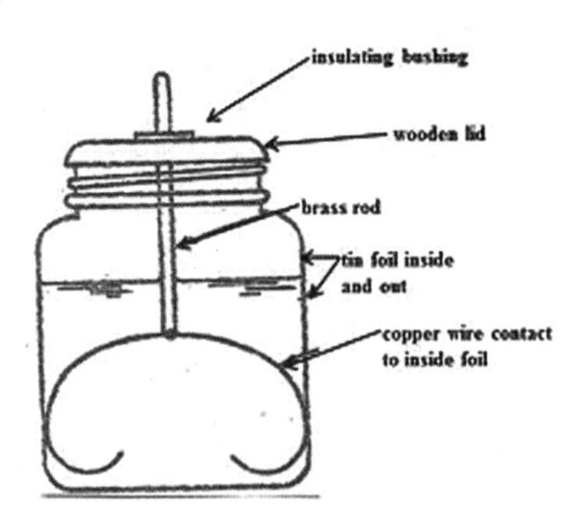

Fig. 2-2. A Leyden jar.

But rather, the kite picked up the electric charge in the air generated by the storm. The electric charge traveled down the kite string to the Leyden jar's central rod and on into the ground below; lightning and other electricity always seek the ground.*

In a Leyden jar, a length of copper wire is connected to the bottom of the brass rod and positioned so the copper touches the interior metal foil. A large negative charge builds up on the internal foil and a positive charge on the exterior foil. The shock from the electrical discharge this storage device produced when Franklin touched it proved to him that electricity had in fact been collected and stored.

Franklin was not the first to fly a kite to collect and store an electric charge; he was not even the first to discover, generate, or store electricity. The first modern scientist to produce a steady flow of electrical current was probably Alessandro Volta. Around 1800, an Italian doctor named Luigi Galvani had found that a frog's leg twitched when it touched two different kinds of metals, which generated a mild current within the leg muscle. Galvani's experiment, which caused movement in a dead limb, may have provided the idea for Mary Shelley's 1818 novel *Frankenstein*, in which lightning animates a dead body assembled from body parts stolen from graveyards.

Volta studied Galvani's findings and concluded that a kind of electrical potential between two metals caused electrical charge to flow through the frog's leg muscle and make it twitch. Volta found that in the presence of this electrical potential, electrical charge can flow through a metal wire like water flowing through a pipe. He used this phenomenon to invent, in 1805, the voltaic pile, a simple device that generated a small but steady electric current.

* We'll see how grounding works in our discussion of the electrical grid on pages 73–74; also see the entry "ground wire" in the glossary at the end of the book.

A friend of Volta's, Luigi Brugnatelli, found that by wiring metal objects to the negative terminal of the voltaic pile, and then dipping the metal in a solution of gold salts, he could cause a thin layer of gold to be deposited on the surface of the metal object. This procedure is called electroplating. As chemist Derek B. Lowe observes, "[in the] early 1880s . . . electricity . . . provided a way to make new chemical reactions happen that had never been seen before."[16] And a new field of science, electrochemistry, was born.

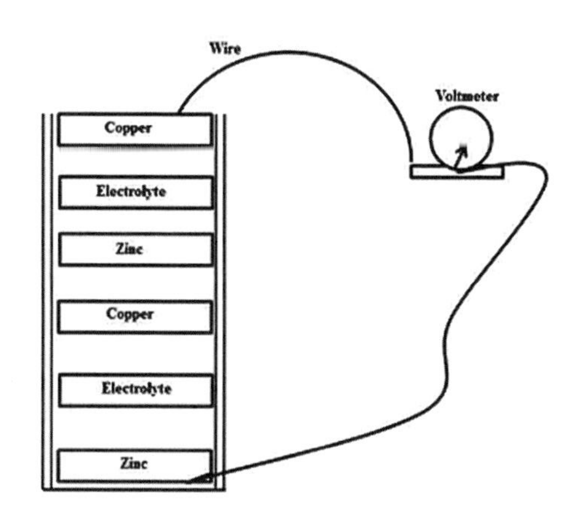

Fig. 2-3. Voltaic pile.

When in high school, I built my own voltaic pile, and it produced enough current to light a small light bulb. The voltaic pile is a stack of three materials, repeated over and over, and always in this order: a zinc disk or square, then an absorbent material soaked in salt water, and a copper disk or square on top. The salt water is an electrolyte, a liquid that can conduct electricity. These metal/electrolyte/metal "sandwiches" are piled atop one another until the pile gets big enough to generate a measurable electric current. I cut the sandwich layers in squares and mounted them in between four vertical wooden posts. By attaching copper wires to the zinc disk at the bottom and to the copper disk at the top, and touching them to a voltmeter, the needle jumped, indicating current was being produced (fig. 2-3).

Although you should not try it at home, you can also detect low voltages with your tongue. When I hold a 9-volt battery that does not have enough power left to run the device from which I took it to the tip of my tongue, I feel a significant electric shock. Again, I warn you not to try this at home.

Different kinds of batteries, like AAA, C, 9-volt, and D batteries, all have different voltages. Scientists have named one of the properties of electricity "electrical potential," also called voltage or electromotive force, which is measured in units of power called volts after Volta. Electrical potential is a measure of the force with which current is sent through the conductor. AAA batteries have an electrical potential of 1.5 volts, transistor radio batteries have an electrical potential of 9 volts, and car batteries have an electrical potential of about 12 volts.

After Volta, many other scientists furthered our knowledge of static and moving electrical charge and its connection to magnetism. These scientists include Michael Faraday, who discovered that moving magnets create electrical current, and Nikola Tesla, who investigated the unique consequences of charge moving back and forth in a wire instead of in just one direction.

But Volta, like Franklin, is also not credited with the discovery of electricity, because the first battery, called the Baghdad Battery, was constructed around 248 BC. It had a carbon rod centered in a clay vase. The rod was surrounded by an electrolyte, which may have been lemon or orange juice, then copper, and then asphaltum, a tar-like substance. The Baghdad Batteries weighed more than 4 pounds each and generated a weak current of around half a volt or less.[17]

How batteries work

The battery was the first reliable source of continuous man-made electric current, and batteries produce direct current (DC) as opposed to an alternating current (AC) such as your local utility power plant generates. Electricity is the flow of electrons through a wire or other conductor, almost always a metal, or through a conductive solution, an electrolyte, such as salt water.[18]

Salt, which is a molecule consisting of a sodium atom bonded to a chlorine atom, dissolves in water. The separated atoms are ions: charged particles. The sodium has a positive charge and the chlorine a negative charge. In a copper wire, electric current is produced by the movement of electrons through the solid copper. In a solution such as salt water, moving ions carry the charges and current. Remember, the definition of current is a charge moving from place to place.

Even though the presence of sodium and chlorine ions makes the salt water extremely conductive, ordinary water is also a conductive liquid, though to a lesser degree. This is why it is hazardous to your health to use electronic equipment, such as a radio or hair dryer, when you are in the bathtub. If the device is plugged into a wall socket and falls into the water, you could be seriously injured or even electrocuted.

Wires carry electric current from the battery to the device it powers. As far as which type of current is best, DC or AC, and why Steinmetz was a champion of AC over DC, we will take up the matter in chapter 4. Given that batteries were invented centuries before generators, clearly DC came first. But how does a battery produce direct current?

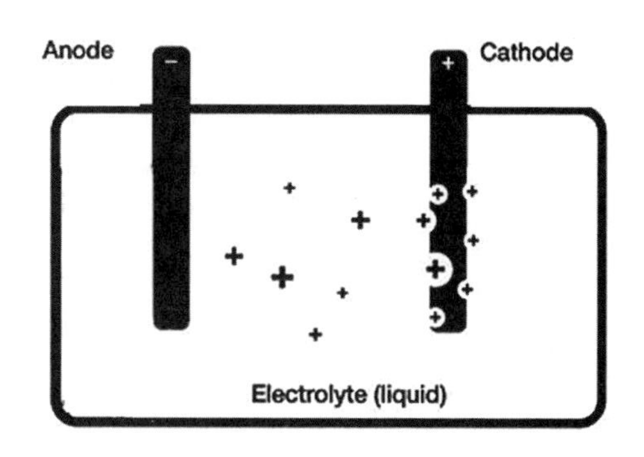

Fig. 2-4. Battery.

To begin with, a battery consists of three components: (1) a negatively charged electrode, the anode; (2) a positively charged electrode, the cathode; and (3) an electrolyte (fig. 2-4).

When both electrodes are connected to an electric circuit, such as the metal within the body of a flashlight (fig. 2-5), chemical reactions in the battery cause electrons to build up at the anode.

The charge within the battery becomes chemically unbalanced, with an excess of electrons at the anode and a deficiency of electrons and a buildup of positive charge at the cathode. So the anode is the part of your battery you see marked "minus" (–) for negative, and the cathode is the terminal marked "plus" (+) for positive.

You know the expression "opposites attract"? Well, it applies to magnetism and electricity as well as people. And just as opposites attract, likes repel. So positive attracts negative. Positive repels positive. And negative repels negative. All the electrons at the anode are negative. So they repel one another. The only place in the battery for an electron to escape the negative charge is the cathode.

However, the electrolyte is a difficult path for the electrons to cross. By connecting the anode and the cathode with a copper wire, which offers less resistance to the traveling electrons than the liquid, the electrolyte keeps the excess electrons from going straight from the anode to the cathode within the battery. When the circuit is "closed," meaning

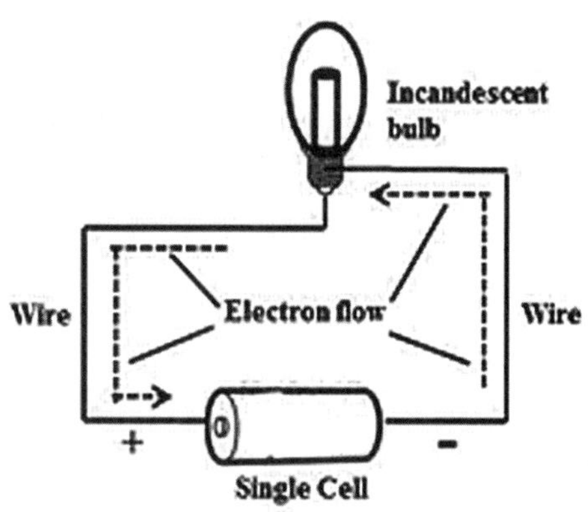

Fig. 2-5. Single-cell flashlight.

a wire connects the cathode and the anode, the electrons move to the cathode through the path of least resistance, the copper wire. The electrons pass through the wire, lighting the light bulb, making the needle on the voltmeter jump, or powering any attached device, along the way. This is one way of describing how electrical potential causes electrons to flow through the circuit.

However, these electrochemical processes change the chemicals in the anode and the cathode, and these electrodes eventually become unable to keep supplying electrons. So there is a limited amount of power available in a battery.

When you recharge a battery, you change the direction of the flow of electrons using another power source, such as jumper cables or a cell phone charger. The electrochemical processes happen in reverse, and the anode and the cathode are restored to their original state and can again provide full power. The most common examples of rechargeable batteries are lead storage cells used for car batteries and the lithium-ion batteries in cell phones and some electric cars such as the Tesla.

For instance, when your car won't start because you left the headlights on, and your battery was drained of all power and is dead, you can run cables from a car in which the battery is working to yours to jump-start your car's battery. Once the motor starts, leaving your engine running continues to recharge your battery, and within a half hour or so the battery is fully charged again.

During a discharge of electricity, the chemical on the anode releases electrons to the negative terminal and ions in the electrolyte through what's called an oxidation reaction. Meanwhile, at the positive terminal, the cathode accepts electrons, completing the circuit for the flow of electrons. The electrolyte is there to put the different chemicals of the anode and the cathode into contact with each another, converting stored chemical energy into useful electrical energy.

These two reactions happen simultaneously. As we stated earlier, an ion is an atom with a charge, and atoms that have given up an electron are positively charged ions. The positive ions transport current through the electrolyte while the electrons flow in the external circuit, and that's what forms a closed circuit and generates an electric current.[19]

If the battery is disposable, such as the kind you buy in a drugstore and put in your flashlight, it generates electricity until it runs out of reactants. But storage batteries, such as those in your cell phone or car, are designed so that electrical energy from an outside source (e.g., the charger that you plug into the wall or jumper cables connected to another vehicle's charged battery) can be applied to the chemical system and reverse its operation, restoring your battery's charge.

Electricity generated over 2,600 years ago

So can we then say that with the building of the first batteries around 248 BC, the discovery of electricity predated Steinmetz and his electrical colleagues by around 2,100 years or so?

In fact, electricity was first observed even earlier than that, when, prior to generating electric current with the first batteries, the ancient Greeks discovered a form of electricity that, unlike current, is not a continuous flow of electrons but rather a quick and sudden discharge of electrons. You know that if you rub your feet on a rug or carpet and then reach out to touch a metal doorknob, a tiny spark jumps between your fingertip and the knob. This is static electricity. If you feel it, the spark is at least 3,000 volts.[20]

Static electricity was discovered more than 2,600 years ago, around 600 BC. Through trading with the Baltic region, the Greeks came to possess amber, a hard yellowish material made from fossilized pine resin. Geologists sometimes find amber with the fossilized remains of prehistoric insects inside it. Well, the ancient Greeks observed that after being rubbed with fur, the amber attracted dried grass and other lightweight objects. Thales of Miletus, a Greek philosopher and scientist who is also credited as one of the first men to trade options contracts, may also have been the first man in Greece to discover that amber, when rubbed with fur, was able to attract small objects to it, though the force of attraction was weak. He believed this attraction property was unique to amber, which turned out to be wrong.[21]

The Greek word for amber is *ήλεκτρον* (*elektron*). And so, the attraction effect was called electric and the energy it produces electricity. Scientists soon found that amber is not the only substance to become electric when rubbed; many other materials become electrified when rubbed with certain other materials.

In 1733, the French chemist Charles François de Cisternay du Fay discovered that there are, in fact, *two* different types of electricity. When amber is rubbed with fur, it acquires so-called resinous electricity, which we now know to be a negative charge. On the other hand, when glass is rubbed with silk, it acquires so-called vitreous electricity, which is a positive charge.

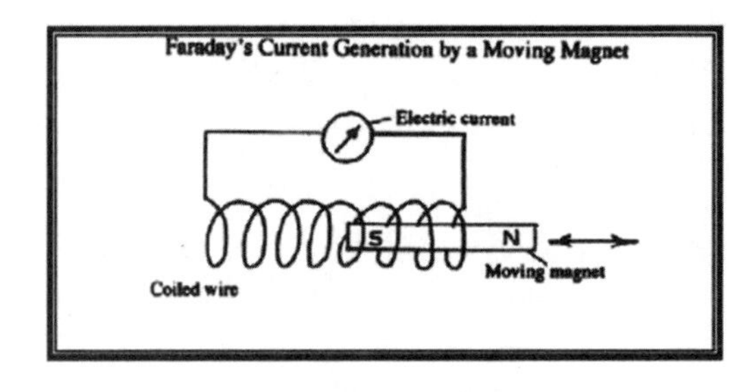

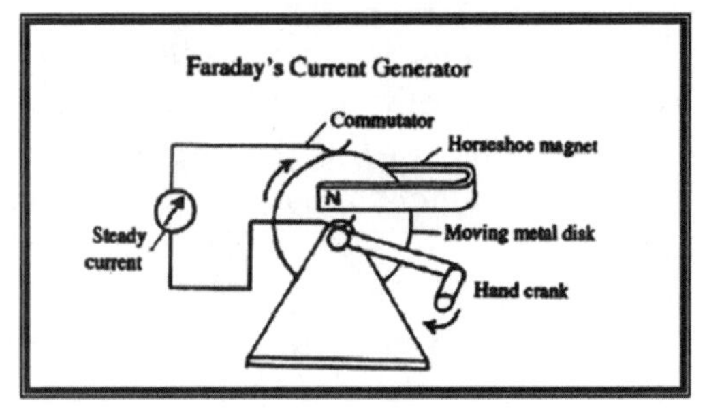

Fig. 2-6. Faraday generated an electric current by moving a magnet through a coiled copper wire. "S" and "N" refer to south and north.

Electricity repels electricity of the same kind, but attracts electricity of the opposite kind.[22] Therefore, when an object with vitreous electricity gets close to another object with resinous electricity, there is a mutual attraction.

The Faraday coil and magnetic induction

Static electricity has a limited number of applications and cannot be used to power light bulbs and other electronic devices. In fact, static electricity can damage sensitive electronic components and devices. In the 1970s, Westinghouse Electric, the company founded by George Westinghouse, produced "Static Havoc," a training video on how to protect electrical equipment from static electricity.

As for batteries, they do not generate enough voltage to power most of our larger modern appliances, let alone supply the electricity needs of a house or other building. You can power your portable CD player and flashlight with batteries but not your air conditioner or refrigerator. Also, batteries generate a form of electricity, direct current (DC), which cannot efficiently be transmitted over long distances.

In 1831, British scientist Michael Faraday (1791–1867)[23] discovered a better way to generate electricity in the form of alternating current (AC), which can be sent over great distances without significant power loss. We will take up the matter of the pros and cons of DC versus AC in greater detail in chapter 4.

Faraday took copper wire and wrapped it to form a coil. He found that when he inserted a magnet through the copper coil, and moved it back and forth, the motion produced electric current. This method of generating electricity is known as magnetic induction (fig. 2-6).

Magnetism is an invisible force that magnets continuously emit without being connected to a power source. Magnetism attracts certain metals, such as iron and steel, called ferromagnetic metals. Oddly, copper, which Faraday used in his magnetic induction apparatus, is not attracted by magnets.

While you cannot see the magnetic lines of force directly, you can observe the pattern they make by placing a piece of wax paper over a bar magnet and then lightly sprinkling iron filings to cover the paper. The iron particles align themselves with the lines of force being continuously radiated by the magnet, as shown in fig. 2-7.

The strength of the magnetic force depends on the distance of the metal from the magnet. That strength is inversely proportional to the square of that distance. Say you double the distance between a magnet and an iron bar. You have multiplied the distance twice. Two squared is 4. The inverse of 4 is ¼. Therefore, doubling the distance between the magnet and the iron reduces the strength of the magnetic force to one-fourth of what it was in the original position.

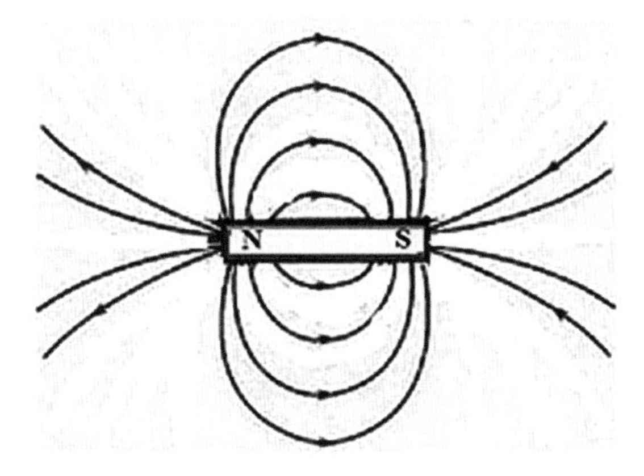

Fig. 2-7. Magnetic lines of force radiating from a bar magnet.

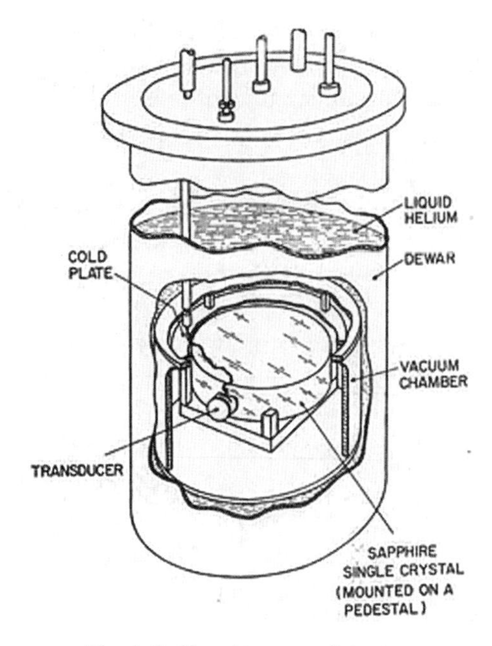

Fig. 2-8. Gravity wave detector.

The earth's magnetic field is also produced by magnetic induction on a large scale: The flowing of liquid metal in the outer core generates electrical currents. The rotation of Earth on its axis spins a large sphere of iron deep in the Earth's core. This generates a magnetic field that surrounds our entire planet. This protective magnetic field prevents the steady stream of particles emitted by the sun, known as the solar wind, from stripping away our atmosphere, thus preserving our ecosphere and the life within it.[24]

Next, Faraday built a prototype device in which the magnet did not move, but in which, by using a hand crank, he rotated a metal coil *around* the magnet. It, too, produced an electric current. Clearly, the movement of a conductive metal within a magnetic field causes electrons to flow, regardless of whether it's the magnet or the conductor being put into motion.

In modern utility power plants, the turbines that generate enough current to provide electric power over large regions are essentially a large-scale version of Faraday's magnetic induction current generator. The turbines spin at high speeds, powering a generator that passes the conductive metal of a large coil through a magnetic field.

While Faraday discovered that moving a wire in a magnetic field makes current flow in a wire, French physicist André-Marie Ampère (1775–1836) also found that a wire *carrying* an electric current *creates* a magnetic field. James Maxwell (1831–1879) analyzed these phenomena mathematically and found that both electrical and magnetic fields are propagated through space as waves traveling at the speed of light, which is 186,000 miles per second in a vacuum.[25]

Einstein posited that gravity also propagated as waves. Gravity waves are generated by mass in motion. But they are extremely weak, so it takes a huge amount of mass to produce gravity waves strong enough for us to measure (fig. 2-8). In 2017, the collision of incredibly massive objects in outer space—two neutron stars—finally produced gravity waves strong enough for scientists to detect.[26]

Inside the power plant

Now, you may wonder: If magnetic induction, a clean energy source, generates electricity, then why do so many utility power plants run on the polluting fossil fuels of oil, natural gas, and coal?

The answer is that burning these fossil fuels generates heat to turn water into steam. It is this steam at high pressure and high temperature that rotates the turbines, which in turn rotate the generators. The steam is directed into a nozzle to produce a high-velocity jet. This fast-moving steam hits turbine rotor blades mounted on a disk and shaft. The momentum of the high-speed steam flowing across the rotor blades produces the force needed to spin the blades, which in turn rotates the shaft (fig. 2-9).[27]

The U.S. Energy Information Administration (EIA) reports that in 2015, fossil fuels generated 81 percent of the energy in the United States,

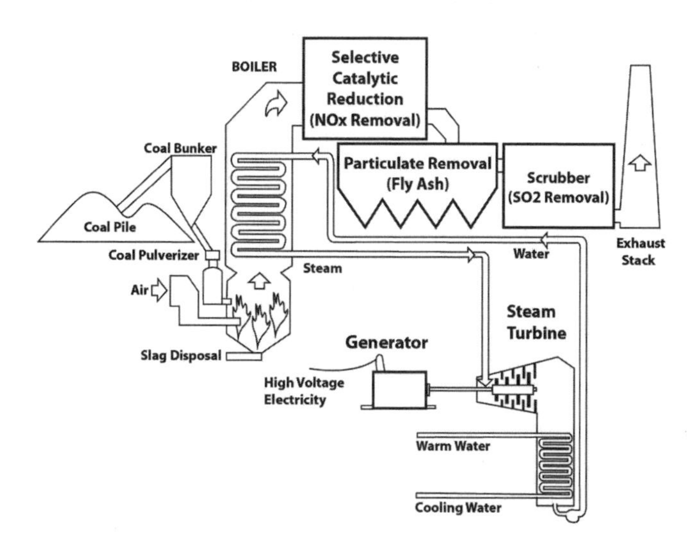

Fig. 2-9. Steam turbine power plant.

with an additional 11 percent generated from renewable energy sources and 9 percent generated from nuclear plants. The EIA also reports energy production in the United States for 2015 totaled 89 quadrillion British thermal units (BTUs). One BTU is the amount of heat needed to increase the temperature of one pound of water—about half a quart—by 1 degree Fahrenheit.

In a large power plant, the generators use magnetic induction as follows: a copper conductor, called the armature, consists of wound coils that do not move. The turbines rotate an electromagnet, so that the magnetic field (not the physical magnet) passes through the armature many times a minute to generate electricity.

Nuclear power plants work in much the same way as coal- or oil-fired plants, only the heat to make stream comes from atomic fuel rather than fossil fuel: the fission reactor generates heat, the heat turns water into steam, the steam rotates the turbine, and the turbine rotates the generator. Many laypeople do not realize that a nuclear power plant does not produce electricity directly from the fission reaction. Rather, the nuclear fission generates heat that is converted to kinetic energy (motion of the turbines) and then electricity.

The advantage of nuclear over coal and oil is this: by avoiding the burning of fossil fuels, the nuclear power plants do not give off polluting gases such as sulfur dioxide, nitrogen oxides, and carbon dioxide. There are, of course, fears and concerns with nuclear power plants, because the fuel, uranium, is radioactive. The radioactivity can make people sick, cause cancer, and even kill them. Children can be born with physical and mental defects as a result of their parents' exposure to excess radiation. Marie Curie, a pioneering scientist in radioactive materials, died from exposure to radiation. Many children born of parents who were in Hiroshima when the atom bomb was dropped on the city during World War II suffered from birth defects caused by radiation that damaged their parents' DNA.

As for power plants that burn natural gas, another widely used fossil fuel, their massive turbines are turned by a hot, high-pressure gas stream, a mixture of natural gas and air. This mixture is heated to 2,000 degrees Fahrenheit. Combustion generates a high-temperature, high-pressure gas that enters the turbine, where it expands. The expanding gas mixture spins the turbine, which in turn rotates the generator, producing electricity by Faraday's induction method.[28]

The problem is that natural gas is highly flammable and combusts easily. For instance, in 2017, a turbine in a natural gas plant in Connecticut caught fire. Because of the intense heat, the fire took three hours to extinguish, and doing so required the turbine to be opened up by workers with bolt cutters and electric saws.[29]

The world is now running out of oil; the supplies of coal and natural gas are also being rapidly depleted. Oil, coal, and natural gas are all nonrenewable fossil fuels, meaning that once they are used up, no more can be made. In addition, burning coal generates greenhouse gases that threaten to heat the earth to catastrophic levels, melting the polar ice caps. Coal's highly polluting nature, combined with coal mining being a dangerous occupation, is causing many areas of the country to switch over from coal to natural gas and renewable energy sources. In Texas, as this book is being written, a group of coal plants generating a total of more than 4,000 megawatts of capacity are being shut down.[30] In Wisconsin, the Pleasant Prairie facility, one of the state's largest coal-fired power plants, was permanently closed after the operation became less profitable and citizens in the neighborhood complained about coal dust.[31] Steinmetz was firmly against burning coal because of the pollution it generates.

Alternative energy systems—wind, water, geothermal, and especially solar—are renewable, in that our civilization and world will in all likelihood come to an end before these energy sources are exhausted. In his 1910 book *The Science of Getting Rich*, Wallace D. Wattles wrote: "Alternative energy forms—such as solar energy and electricity produced through wind and other natural forces—are businesses that are still in their infancy and have great potential."

Renewable energy is clean and nonpolluting. For these reasons, the United States and many other countries are accelerating the effort to rely less on fossil and nuclear fuels and more on renewable energy. Of these renewable energy sources, Steinmetz was an advocate of water power over all the others.

Interestingly, whether the electricity is produced by burning a lump of coal or converting a stream of sunlight into current, the electrical grid system pioneered by Steinmetz, Westinghouse, and Tesla will still be here to carry it to our homes and offices.

Home solar installations may lighten the utility system's load, but large-scale central power plants will continue to provide the bulk of our energy for the foreseeable future. In fact, homeowners installing solar panels that generate more power than the house consumes can now sell the excess electricity back to their utility company.

In a recent survey of 500 medium and large businesses by Zpryme, nearly nine out of ten companies said they plan to implement energy efficiency. Eight out of ten said they are likely to look into renewable energy. The number one motivation for these businesses to pursue energy efficiency is cost savings.[32] As for consumers, 56 percent of Americans surveyed said it is important for their home to be energy efficient.[33] In 2017, Georgetown, Texas (population 67,000) became the largest city in the United States to be powered entirely by renewable energy.[34]

Solar energy

Our planet is about 92 million miles from the sun, which means it takes light around eight minutes to make the trip from our sun to the earth. The side of the planet facing the sun gets more solar energy in one hour than the entire human race uses in a year. Scientists have measured the energy pouring down on the earth from the noonday sun and found it to be a constant 1.97 calories per square centimeter per minute. A calorie is equal to the amount of heat needed to raise the temperature of one gram of water by 1 degree Celsius. The sun has been sending this energy to our planet throughout all of history . . . and will continue to do so for billions of years.[35] The energy produced by our sun in just five billionths of a second is equivalent to the entire energy output of all people on Earth combined for an entire year.[36]

There are several methods for converting sunlight into usable energy. One of the simplest is to use a dish or reflector to capture and concentrate the sun's rays. In a "parabolic-trough" system, a series of long, rectangular, curved mirrors are tilted toward the sun. The mirrors focus the sunlight on a pipe that runs down the center of the trough, heating oil flowing through the pipe. The heat from the oil is transferred to water, boiling the water to produce steam. The steam is used to generate electricity. The cooled oil is returned to the pipe so it can be reheated again by the solar rays.

Another technology, the solar cell, converts the sun's rays directly into electricity through the photovoltaic method, based on the photoelectric effect discovered by Einstein, which we will discuss later on in the book (see page 111). The first solar cell, introduced by Bell Telephone Laboratories in 1954, was made from a small silicon wafer.

In a solar cell, photons strike the silicon, causing electrons in the material to move. The movement results in uneven distribution of electrons in

Fig. 2-10. Home powered with rooftop solar energy panels.

the silicon. If you connect a wire between the two sides of the cell, electrons flow from where the concentration is high (the negative pole) to where there are fewer electrons (the positive pole), creating an electric current. The photovoltaic process does not deplete the silicon, so the solar cell can generate current almost indefinitely without wearing out. Even though silicon is one of Earth's most plentiful elements, photovoltaic (PV) cells can be expensive to manufacture. This is because PV cells require an extremely high-purity silicon that must be refined through extensive processing.

However, the high cost of oil—combined with newer, more efficient solar conversion technologies—has made solar energy a more cost-competitive alternative to fossil and atomic fuels than ever before. Hawaii, for instance, has announced a plan to have all of its energy produced by renewable energy sources by 2045.[37]

In New York State, Delaware River Solar and Ampion recently built a 2.7 megawatt solar array serving over 350 customers.[38] California, which is the fifth largest economy in the world, has mandated that all new homes built after 2020 have solar panels installed.[39] Worldwide, solar power has doubled in capacity every two years for the last two decades.[40]

Fig. 2-11. Modern windmills.

Wind power

For wind power, spinning windmill blades generate electric current. With zero emissions, wind farms are far cleaner for the environment than coal, gas, or oil. Wind turbines generate no pollution and do not contribute to global warming.

In modern windmills (fig. 2-11), large spinning blades in the windmill's turbine capture the kinetic energy of the moving air (the wind). The kinetic energy is transferred to rotors that generate electric current.

Just a few decades ago, wind systems were inefficient. For instance, in the 1980s, it cost 40 cents per kilowatt to generate electricity from wind turbines—about 10 times the cost per kilowatt of electricity produced by burning fossil fuel back then. As technology improved and crude oil prices climbed higher, wind power became increasingly cost-competitive on a per kilowatt basis with oil, natural gas, and coal. Today the average U.S. consumer pays about 12 cents per kilowatt-hour (kWh) for electricity. The majority of that expense is utility overhead, while the actual cost of electricity generated is 2 to 4 cents per kWh.[41] Wind power, at around 8 cents per kWh, is not quote competitive with that yet, but it is getting closer and closer.[42]

Only 1 percent of the United States' electricity is generated through wind power. The Department of Energy (DOE) has set a national objective of generating 20 percent of the country's electricity from wind by 2030. Bloomberg has forecast that the United States will have a total installed wind energy capacity of 3 or 4 gigawatts by 2030.[43]

Hydroelectric energy

In hydroelectric plants, falling water spins the turbine. In the early days of America, the kinetic energy of falling or flowing water was converted into mechanical energy mainly by small wooden waterwheels, usually to turn simple machinery in mills for grinding wheat into flour or textile manufacturing. Alexander Hamilton located his first industrial plant in Paterson, New Jersey (where I grew up), at the base of the Great Falls to take advantage of the power of this huge waterfall. The Great Falls are the second-biggest waterfall on the East Coast of the United States. Niagara Falls is the largest, and it also had a spillway built to turn a waterwheel for a mill. Perhaps the most famous U.S. source of hydroelectric power is Hoover Dam on the border of Nevada and Arizona. Built with 6.6 million tons of concrete, the dam is 726 feet high, has 17 turbines, and provides electricity to 1.3 million homes.[44]

Unfortunately, waterwheels can be located only where a rapidly flowing stream exists or can be made, through damming, to produce a flow of water with enough kinetic energy to turn the wheel. However, there's one place on Earth where water is always flowing with an incalculable amount of kinetic energy: the ocean. If you've ever gone swimming in the ocean and been pounded by waves or dragged out to sea by the undertow, you've felt this force. Now engineers at several companies are building devices that can convert the force of ocean waves, swells, and tides into usable energy.

One such device, the "Seadog" wave pump, uses the mechanical energy of ocean swells to pump seawater to land-based hydroelectric turbines. The water moves through the turbines, spinning the blades to generate electricity, and then is returned to the sea. But the turbines still have to be near the shore; you can't run a Seadog wave pump in Arizona.

Another technology for generating electricity from ocean waves consists of buoys tied to generators. As waves pass, the buoys bob up and down, creating a flow of hydraulic fluid. The fluid flow powers an electrical generator sealed in waterproof housing on the ocean floor. A cable carries the electricity from the generator to a transformer on the shore.

The buoys are 15 feet in diameter and 40 feet long. Each can generate up to 50 kilowatts of electricity, enough to light up 50 homes. Naturally, the areas most likely to benefit from this technology are towns and cities on or near the ocean.

In 2003, a wave-based power generation system began providing electricity to homes on the Arctic tip of Norway. Tidal currents in the Kvalsund sea channel turn the 33-foot-long blades of a turbine, which is bolted to the seafloor. During the 12 hours each day when the tides rise and fall, they send water in and out of the channel at a speed of 8 feet per second. The Kvalsund turbine system costs about $11 million. It generates approximately 700,000 kilowatt hours of energy a year, enough to light and heat about 30 homes.

Steinmetz was extremely enthusiastic about hydroelectric power, which he referred to as "white coal." He believed, as it turns out too optimistically, that eventually hydroelectric power would be able to meet all the power demands of the entire continental United States.[45]

Geothermal energy

The problem with the conversion of geothermal energy into electrical energy is that it can only work in areas where there is sufficient heat in the ground or water. Unfortunately, only a few places in the world have sufficient geothermal activity to make geothermal energy a significant source of alternate energy, and therefore it provides a small percentage of the earth's total energy production.

Australia has large reserves of geothermal energy in granite buried within 6 to 10 miles of the surface. One cubic kilometer of hot granite at 250 degrees Celsius holds geothermal energy equivalent to 40 million barrels of oil. Iceland also has rich geothermal fields, as well as many fast-flowing rivers for hydroelectric power generation. As a result, two-thirds of Iceland's energy is produced from alternate (renewable) sources, mainly water and geothermal. In the United States, the most geothermal activity is in Yellowstone National Park in Wyoming, which is significant mainly as a national park and tourist attraction and not an energy source. However, in New England, a bubble of hot rock, a new potential source of geothermal energy, is now rising under the northern Appalachian Mountains.[46]

Biomass energy

Another source of alternate energy is biomass fuel: burning garbage and other organic material. Ample energy produced through combustion of any flammable material, including trash. The challenge in biomass energy is twofold. The first is efficiency: burning the biomass in such a way that a large portion of the energy is transformed into usable power. The second is pollution: not releasing or creating harmful emissions as the biomass is burned.

One type of biomass energy generator is "gasification" technology. Gasification systems produce energy by burning carbon-based waste products such as biowaste, agricultural waste, and municipal sewage. The biomass is incinerated in such a way that burning it produces "synthesis gas," or syngas for short. Syngas contains hydrogen, carbon monoxide, carbon dioxide, and nitrogen. It has a high-energy content and can be compressed and stored for later use.

How does gasification work? The organic material in the waste reacts with steam and oxygen at high temperature and pressure, which chemically converts it to syngas. The extreme temperatures incinerate the nonorganic materials in the garbage, such as metals and plastics, into ash. The ash is inert and has a variety of uses in the construction and building industries. The synthesis gas produced by the gasification process can be used to drive a turbine engine driving an electric generator to produce electricity. Or, it can be catalytically converted to produce ethanol, natural gas, or anhydrous ammonia.

Another technology for converting biomass—in this case, wastewater—into power consists of a fuel cell with graphite electrodes and a catalyst membrane made of carbon, plastic, and platinum. The wastewater is piped into the fuel cell. Microbes within the wastewater generate free electrons as their enzymes break down sugars, proteins, and fats. During the process, up to 78 percent of the waste products in the water are removed. The free electrons, meanwhile, produce 10 to 50 milliwatts of power per square meter of electrode surface.

A biomass technology used by Duke Energy in North Carolina converts methane from pig poop into a gas that can be burned at electric power plants.[47] In Maine, a pair of biomass plants burn leftover wood from forestry work and sawdust from mills to make electricity.[48]

Enter Steinmetz and the grid . . .

The most efficient way to transmit and distribute energy, especially over long distances, is as electricity. But electricity produced by generators cannot be stored efficiently. As a result, electricity produced by power plants must be delivered and used as soon as it is generated. To accomplish this quick distribution and consumption requires a large, interconnected network capable of carrying a high load. This network was in fact built and is the electric utility power grid, sometimes referred to as "the grid," that crisscrosses the United States today. An article in *Smithsonian* magazine observes, "Taken together, the generation and distribution of electric power in the United States is an astonishingly complex undertaking."[49]

In an AC circuit, a cycle is defined as one complete revolution of the generator coil, and the alternating current reverses direction twice per cycle. In a 60 hertz (Hz) system, which operates at 60 cycles per second, the current reverses direction 120 times per second. At this rapid rate, the flow of current registers as continuous on a voltmeter, and the changes in direction are not visible to the naked eye.[50]

In U.S. power plants—those currently powered mainly by fossil fuels, nuclear, and to a lesser degree water— the generators spin, and the magnetic field surrounding the magnet passes through the metal armature at a speed to produce 60 Hz of alternating current.

Many other countries around the world use 50 Hz current, which is not compatible with your laptop computer, mobile phone, and other electronic devices bought in the United States. If you plug a 60 Hz device into a 50 Hz outlet, it can damage or destroy your equipment. That's why people who travel overseas must carry a universal power adapter. This is a small unit that plugs into outlets producing current of 50 Hz or some other frequency, and converts that power to 60 Hz or, if you are not from the United States whatever other hertz you need. Get a universal adapter, which is inexpensive, before you travel out of the country!

Large power plants today produce on average 20,000 volts of electric current, while smaller generators have outputs of 400 to 6,000 volts. But let's say we put up large power plants all over the country, as in fact we have. We still have the problem of getting the electricity they produce to the homes and factories that need it. The solution is to build, operate, and maintain the grid, the nationwide network of cables that carry electric current from the utility power plant to your house or workplace.

The use of electricity to power technology was predicted by Albert Robida in his 1890 novel *The Electric Life*, which features the "telephonoscope"—a combination telephone and TV set—as well as an electric-powered rapid transit system, submarines, and aircraft.[51]

In Theodor Herzl's 1902 novel *Altneuland*, Altneuland is a utopian paradise established by a colony of Jews on a remote Pacific Island. The colonists have created advanced technology, including electric trains, boats, and cars, all powered by electricity, which was fairly new and not widespread in usage at the time of publication.

In Otfrid von Hanstein's 1928 science fiction story "Electropolis," a city of the same name occupies half a million square miles in the Australian desert; electrical charges from huge towers trigger rain, making the desert bloom.

As noted, most power plants generate electricity by burning coal, oil, or natural gas, though the United States has 61 additional commercially operating nuclear power plants.[52] There are also 1,756 government and private hydroelectric power plants in the United States, the majority of which are relatively small.[53]

What Steinmetz brought to the world was not a better method of generating power—others before and after him took care of that—but a better way of getting that power to the people and companies that need and use it via the electrical grid.

Here's how the grid works: From major power plants, the electric power is sent over wires to smaller plants called substations. Each power station is connected to dozens of substations and thousands of lines and poles. The grid's substations are located relatively close to where the electricity will be used. Power lines carry the electricity from the power plant first to the substations and then to tall metal towers. From those towers, electricity moves through large power lines to smaller ones and into buildings. Then individual wiring systems within the building distribute the electricity to where it is needed.

Two properties describe the electricity involved in the transmission of power over the grid: voltage and amperage. As stated earlier, voltage is the pressure or force that drives current—the electrons—through the wire. Amperage is the amount of energy contained in the current. Voltage is directly proportional to amperage. That means the higher the voltage, the higher the amperage.

High amperage poses a problem. When the electrical energy is too great, much of it is lost as heat, reducing the amount of current that ultimately gets delivered to the utility customer. This loss is related to a phenomenon known as hysteresis, which was discovered by Charles Steinmetz and will be taken up in chapter 3. The solution is to transmit power with lower amperage and greater voltage through a conductor with sufficient resistance to reduce amperage, as resistance and amperage are inversely proportional.

Fig. 2-12. U.S. continental electricity grid.

But voltage that is too high is also a problem. Reason: If there is enough voltage or force to transmit a current through several hundred miles of wire, this large voltage may cause the current to jump off the transmission line in places where insulation is thin, worn, or missing.

The continental United States is divided into three major electric grids: Eastern Interconnection, Western Interconnection, and Texas Interconnection (fig. 2-12). Combined, they are the National Power Grid. The continental grid consists of a vast network of power plants and transformers connected by more than 450,000 miles of high-voltage transmission lines.[54] Separate electrical grids distribute power in Alaska and Hawaii. Puerto Rico is also on a separate grid, which was temporarily disabled in 2017 by Hurricane Irma.

In addition to the three major grids in the continental United States, there are numerous "microgrids." A microgrid is a small network of electricity users with a local source of supply. The microgrid is usually connected to the National Power Grid but can operate independently. Because of the increased demand for clean, reliable power in remote regions and developed countries, the global microgrid market is forecast to grow from $9.8 billion in 2013 to $35.1 billion in 2020.[55]

The "distributed grid" moves electricity from a more centralized grid in which one-way power flows from a central power plant to customers into a more decentralized, two-way grid in which electricity is increasingly generated by smaller-scale power sources. The power is either used directly by commercial, industrial, and residential customers or shared among energy consumers. In 2017, more than half of commercial and industrial customers said the distributed grid's importance will increase within the next two years.[56]

When a smaller regional power grid operates on its own, it can draw only on its own resources. But increasingly, microgrids are being linked to major grids. Connecting the smaller grids has given more people access to reliable and sufficient electricity. This is especially important during periods of high demand in a particular area. For example, the interconnected resources of the entire Eastern Power Grid might supply power to Alabama for a few months during a very hot summer. Then, during a severe winter, grids in Florida might send power to cold New England states. However, once the grids are connected, the failure of one can lead to the failure of them all.

Electricity can be very dangerous, so power plants are set to shut down in times of trouble. The shutdowns can prevent massive fires and other damage. But the shutdowns also cause people in that area to temporarily lose power.

You may have seen large electrical equipment surrounded by a fence with a sign warning people not to enter the area because of risk of severe electric shock, which could easily cause death. These power substations are equipped with transformers. A transformer is a piece of electrical equipment that increases the force at which electrons move. As we discussed, this force is called voltage. Increasing the voltage means more electricity can be transmitted over a wire or cable.[57]

Some transformers, called step-up transformers, boost the voltage from the power plant to between 155,000 and 765,000 volts. Doing so reduces line losses and enables transmission of the electricity over a distance of about 300 miles. Then, transformers on poles or underground power lines at the receiving end reduce the current to lower voltages, so it can be sent to homes safely. These local transformers are called step-down transformers, because they take the voltage intensity several steps below the peak level of the transmitted power. After the electricity moves from the transformers, it travels through electrical wires and into homes and businesses. Before entering homes and other buildings, the electrical power is reduced even more. Small step-down transformers on top of electrical poles bring the power down to safer levels.

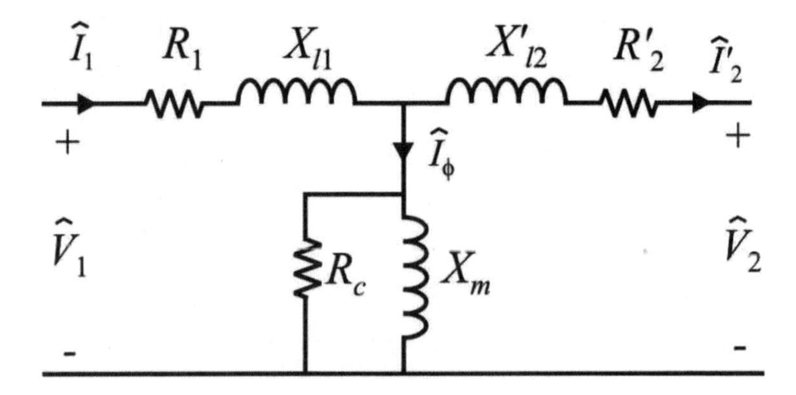

Fig. 2-13. Steinmetz's design for a transformer, where R is resistance, V is voltage, and I is current.

As we shall see later in the book, this grid exists in its modern form based on the pioneering innovations of Steinmetz, Tesla, and Westinghouse. As for the light they shine on our towns and cities, we owe that to Thomas Edison and other scientists who developed light bulbs. You can debate who made the greatest contribution to alternating current transmission; in this book, I make the case that Steinmetz arguably worked out the most rigorous mathematics that today govern AC distribution.

Though many great scientists were born in the United States, Steinmetz, the father of the AC power distribution grid, was not. And we are lucky he immigrated to America and could do science here, as he was almost denied entry into the country when he arrived.

3

The Mathematics of Electricity

A natural aptitude for or strong interest in science can enable a student to become a reasonably competent or in some cases even a highly accomplished scientist. Similarly, mechanical ability can lead to a rewarding career as an engineer or an inventor. However, many educators and professionals agree that students and practitioners with a mastery of higher mathematics have a big advantage over those who are less at home with numbers and equations.

As Dr. Daniel Kleitman of MIT notes, "The development of calculus and its applications to physics and engineering is probably the most significant factor in the development of modern science beyond where it was in the days of Archimedes. And this was responsible for the industrial revolution and everything that has followed from it, including almost all the major advances of the last few centuries."[1]

"The whole world is made up of [math] problems," says Steve Cutchen, U.S. Chemical Safety Board Investigator. "If you're not good at math, you're going to get taken advantage of your entire life."[2]

Professional engineer Conrad Gamble says, "An engineer without good math competency would be similar to driving car without a set of working windshield wipers. This person would be limited to tasks within their capabilities. That could certainly limit the opportunities available for the engineer."[3]

For instance, one of my college roommates, who entered the university planning to major in electrical engineering (EE), was a whiz at electronics. John could perform miracles with a pair of pliers, a coil of wire, and a roll of duct tape. But then John took freshman calculus and found the material utterly impenetrable. Unable to pass integral calculus, he could not go forward with his studies and obtain the EE degree he had counted on, and so entry into the profession was denied him.

Steinmetz, on the other hand, had a natural talent for mathematics. Ronald Kline, an instructor at Cornell University, wrote, "Steinmetz helped make electrical engineering more scientific by introducing more rigorous mathematics into the profession."[4] For example, Steinmetz became a major contributor to working out the advanced equations governing the transmission of alternating current. His foremost contribution is perhaps developing the law of hysteresis, which determines power losses in electric transmissions.

In his book *Steinmetz: Maker of Lightning*, Sigmund Lavine wrote, "His skill with figures was a constant source of bewilderment to everyone associated with him, and as he solved one problem after another, his fellow workers became more firmly convinced that [Steinmetz] could do anything with mathematics—from figuring out the proper ratio in a set of gears to squaring the circle."[5]

Steinmetz was indeed a first-rate mathematician; his books are filled with endless equations beyond the understanding of most laypeople, including, for the most part, me, despite my training in the field. An article on the Edison Tech Center website notes, "Steinmetz was the first person to understand AC power from a mathematical point of view"—just as Maxwell was the first to master the mathematics of electrical and magnetic fields.

Hysteresis

Remember the toy called the Slinky? It was coiled metal or plastic wire. You placed it on a step and gave it a gentle push (fig. 3-1). The front of the Slinky curved and landed on the step below. Then the bottom end followed. The momentum caused a repeat of the action, causing the Slinky to "walk" down the stairs. Notice that when you push the Slinky, though the front end moves and takes the next step down right away, the rear seems to lag for a second before it too springs down to the step below.

The Slinky is a demonstration of the principle of hysteresis, which occurs throughout nature, both in the visible mechanical world and the invisible world of electromagnetic fields. Hysteresis is the phenomenon in which the value of a physical property lags behind changes in the effect or source causing it. In the case of the Slinky, it's the lag between pushing the Slinky forward and when the rear end leaps into action to follow the front.

Fig. 3-1. Slinky.

When the front slinks down and lands on the next step, the toy is virtually motionless for perhaps a second or less, and then, after that brief delay, the rear end obediently follows the front downward, flips over it, and lands on the step below it.

When magnetism alternates or reverses direction back and forth, it consumes energy, generating heat and losing efficiency in the production of electric current. In electromagnetism, the loss of power caused by the alternating magnetism is the hysteresis effect. One reason hysteresis takes place in pumps and motors is that the iron in motors resists being magnetized or demagnetized.[6]

When the magnetic field is shut off, you would think the flow of electricity it produces would stop as well. But the current declines only gradually, albeit often quickly. Still, the effect—the drop of current to zero—lags behind the cause of shutting the power (fig. 3-2).

Steinmetz wrote the first papers published in technical journals about hysteresis and figured out the laws governing hysteresis in electrical circuits. This better understanding of the law of hysteresis meant electrical circuits and transmission lines more efficiently accommodated the behavior of the current they carried, with less power loss, which enabled

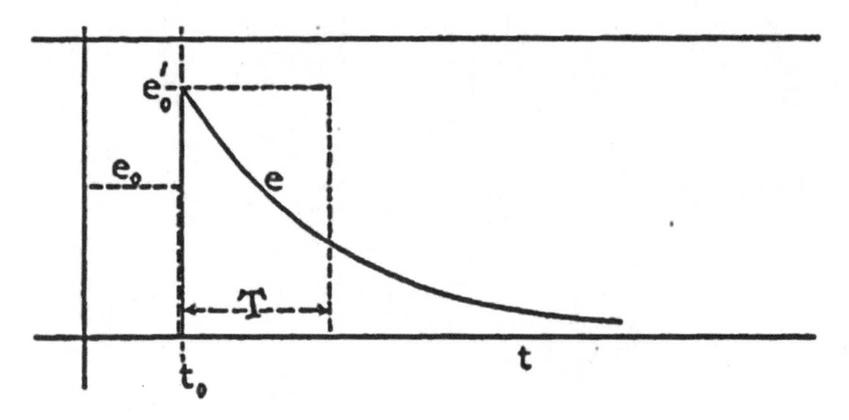

Fig. 3-2. Hysteresis in an electrical circuit shows the current ("e" above) does not instantly stop when you switch it off. Instead there is a delay ("t" equals time) between shutting off the circuit and cessation of electron flow through the wire.[7]

current to travel over greater distances. Result: an important step forward in achieving optimal power generation and conveyance in the utility grid.

In electricity, an example of hysteresis is seen in magnetic induction, where passing a copper wire or other conductor through a magnetic field generates electric current. Hysteresis is *always* present in magnetic induction, meaning that there is always a delay, however brief, between the start of the flow of electric current and when the conductor passes through the magnetic field.

Here's another electrical example of hysteresis. Say you connect a 12-volt relay to a power supply. Relays are switches that open and close to permit or halt the flow of current in a circuit. In the open position, current flow should stop. And it does. But not instantly. To start, you first gradually increase the voltage from zero to 12 volts. The relay switch activates when the current hits 11 volts and above. So naturally you would think that if you go in the opposite direction, lowering the voltage, when you reach the point where the voltage drops *below* 11, the relay would shut off. But this is not the case. Although it seems to defy logic, the relay only deactivates after the current drops below 9 volts. The voltage lag of the relay shutting off between the 11-volt activation threshold and the 9-volt deactivation is hysteresis.

The Slinky is one example of mechanical hysteresis. Another example occurs in machine shops using computer-controlled lathes to work with metal. Lathes have specialized types of clamps called chucks. In the operation of a lathe, these chucks spin, and their direction is frequently slowed or reversed. When that happens, the machine's clamping pressure tends to increase. This increased pressure can distort or damage the metal being worked, especially parts with thin walls or made of soft metal. As a result, some chucks are built to compensate for hysteresis with counterweights inside the chuck body. These weights exert pressure in a direction opposite the deceleration, because of their forward momentum. This in turn stops the hysteresis from damaging the part.[8]

Steinmetz's mathematical analysis of hysteresis has usefulness beyond alternating current transmission. Hysteresis is a phenomenon seen in the behavior of many different types of magnetic materials. A partial list of materials and equipment in which hysteresis takes place includes permanent magnets, magnetic recordings, data storage, electric motors, transformers, and many other electronic devices.[9]

Magnetism is the driving or causative force in the production of electricity through induction. The electrons in magnetic materials are aligned in one direction, or "polarized," so there is a positive pole and a negative pole. Therefore, magnetic materials continuously radiate magnetic fields. The maintenance of a magnetic field does not require power for its maintenance. In fact, if you rub a ferrous metal (any metal containing iron) with a magnet in the same direction for a time, it will cause the electrons in that metal to line up and become polarized, so that the metal, too, generates a magnetic field, though the magnetism in the metal is temporary and weaker than an actual magnet.[10]

Electricity, on the other hand, requires power to generate. This can be achieved via magnetic induction, and it takes kinetic energy, such as turning a turbine, to create the moving magnetic field that produces the electrical energy. Or, as in a battery, chemical energy is converted to electrical energy.

The strength of magnetic attraction is called magnetic intensity, and the size of the magnetic field passing through a given area is the magnetic flux. The movement of the armature or coil passing through the magnetic field requires energy, but the steady presence of the magnetic flux it interacts with does not. In 1826, the German physicist Georg Ohm (1789–1854) suggested that the intensity of current in a circuit was governed by the formula $E = IR$, where E is the electromotive force (voltage), I is the current

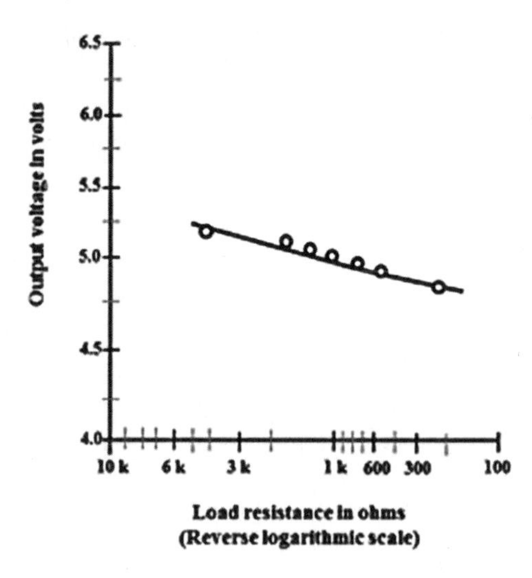

Fig. 3-3. Voltage versus resistance.[12]

intensity, and R is resistance—so named because it resisted or opposed the flow of current.

According to the E = IR formula, called Ohm's law, the greater the resistance (R), the lower the intensity of the current (I). Three primary factors control resistance: wire or cable length, thickness, and material. The thicker and longer the cable, the greater the resistance, which is why transmitting current over long distances without huge power losses is so problematic—a problem Steinmetz was instrumental in solving. Also, as shown in figure 3-3, resistance varies depending on the material the cable or wire is made of. For instance, tungsten has a resistance three times greater than copper.[11]

As Einstein showed, energy is always conserved; it is never created or destroyed. So when resistance lowers the intensity of electrical current, what happens—where does the vanishing energy go? Answer: The energy created by this resistance is stored as potential energy in the magnetic flux. If the magnetic flux is decreased, this energy is dissipated by the change in the flux. In a phenomenon called molecular magnetic friction, the energy is converted into thermal energy, or heat. When a magnetic flux is periodically changed by an alternating or pulsating current, this dissipation of energy by molecular friction occurs during each magnetic cycle, lagging behind the magnetizing force, and hysteresis results, giving rise to a "hysteresis loop" (fig. 3-4).

To be clear, hysteresis and molecular magnetic friction are not quite the same thing, though they are closely related. The hysteresis loop measures the molecular magnetic friction only when energy is supplied to the electrical circuit by the magnetizing current—in other words, when electricity is produced through magnetic induction.

So what does that mean? As mentioned, it takes mechanical work to move the armature or wire coils through the magnetic field. With that action, the hysteresis loop enlarges. The expansion is caused by both the energy dissipated by molecular magnetic friction and the energy converted into mechanical work.

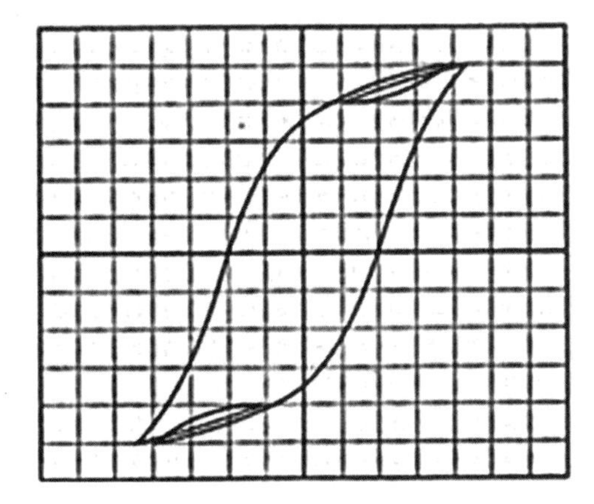

Fig. 3-4. Hysteresis loop.[13]

Mechanical energy can also be supplied to the magnetic circuit by vibrating it mechanically, as is done in synchronous motors, applications of which include position control, fans, flowers, and to drive axles in off-road vehicles. The vibration of the magnetic circuit causes the hysteresis loop to collapse; its area becomes equal to the molecular magnetic friction minus the mechanical energy absorbed.

The hysteresis cycle is largely independent of the frequency of the alternating current, no matter how rapid the cycles. The only noticeable effect at extremely rapid cycles is that it generates "eddy" current. Eddies are small local loops of current within the conductor. These eddies create resistance blocking the main flow of current through the wire. They also give rise to small local magnetic fields that weaken the main magnetic field used to produce magnetic induction.

On the other hand, extremely low-frequency cycles can alter the hysteresis, making it erratic and irregular through a phenomenon known as "magnetic creepage." Typically the effect on AC circuits is negligible.

In modern science, the credit for an invention or discovery traditionally goes to the scientist who publishes his or her results in a recognized scientific or technical journal. In 1892, Steinmetz wrote and published the first major technical article on hysteresis, "On the Law of Hysteresis."[14]

Now, if hysteresis is the lag between an applied force and the action or reaction that follows, could the opposite phenomenon also occur—that is, an action happening *before* the causative force or agent is actually applied? Of course not, but in 1948 Isaac Asimov wrote "The Endochronic Properties of Resubliminated Thiotimoline," a spoof article claiming he had done so in an experiment.

Asimov noticed that when you drop a crystal of a compound into a solution already saturated with that same compound, the entire solution solidifies into crystals of that compound almost instantaneously as soon as one crystal hits the liquid. He thought: What if the crystallization of the solution took place a fraction of a second before the crystal was added to the saturated solution?

This would be the opposite of hysteresis, with the reaction taking place before the driving force, rather than after the force is applied. Of course, this reverse hysteresis was purely imaginary; Asimov knew it was a fiction and wrote the spoof paper for his own amusement, which was natural given he was not only a chemist but also a science fiction writer.

AC steady-state circuit theory

Hysteresis was one of two major discoveries made by Steinmetz and other top electrical scientists of his day. The other was that not all circuits behave in a similar manner: in some, the circuit characteristics remain constant, while in others, they vary unpredictably.

Steinmetz and others suggested that there are two classes of AC electric power: steady state and transient. In the steady state, all current flows and voltages are constant as long as the condition of the circuit remains unchanged. In the transient state, changes in the condition of the circuit cause changes in the current and voltage. When the circuit conditions change, the change in voltage and current does not happen instantaneously. Rather, the resultant rise or fall in voltage and current is delayed, meaning the effect lags behind the cause. This is the phenomenon of hysteresis described in the previous section.

Transients can be a problem. For instance, say you are using an electric pump to fill a basin with water. When the basin is almost full, you shut the pump off. But instead of stopping instantly, hysteresis causes the pump to continue to operate after it is turned off. So the water continues to flow for a time, and the basin may overflow.

As discussed with our earlier example of the Slinky toy, hysteresis applies not only to electrical energy but also to other forms of energy, such as mechanical and heat. For example, the engines of a cruise ship propel the giant vessel through the ocean. But when the engines are stopped, the boat does not stop instantly. Instead, the motion slows gradually (fig. 3-5), and it can be some time after turning off the propulsion before the boat finally comes to a full stop.[15]

Fig. 3-5. Deceleration of a ship.

AC steady state is the term used to describe alternating current circuits in which the major components—including resistors, capacitors, and inductors—all operate in a linear fashion. Because of the linear operation of its components, the voltages and currents in the circuit can be calculated using differential equations, a calculus measuring the rate of change of quantities.

In steady-state circuits, the variables of the characteristics of the electric current—which include voltage, resistance, and capacitance—are stable over time as they propagate over the power line. The voltage and current sources generate wavelengths in the pattern of a consistent, unchanging sine wave, with constant frequency, amplitude, and phase—and no transients. In steady-state AC circuits, two transverse electromagnetic waves travel along the power cable in opposite directions, the direction switching back and forth 60 times a second in standard 60 Hz AC current.

Prior to Steinmetz, electrical engineers had to calculate voltage and current in steady-state AC circuits with difficult methods dependent on high-level calculus. Then Steinmetz went to work on the problem and revolutionized AC circuit theory and analysis. In a landmark paper, "Complex Quantities and Their Use in Electrical Engineering," presented at a July 1893 meeting of the American Institute of Electrical Engineers (AIEE), Steinmetz showed how to simplify these difficult calculations to a problem that could be solved with basic algebra, greatly simplifying the mathematics of alternating current. As a result, engineers were able to design circuitry for a wider range of applications more quickly and easily than they could using the more complex calculus.

AC transient theory

In comparison with steady-state circuits, transient circuits are characterized by surges and oscillations in waves that are produced by disturbances within the circuit.[16]

Transient analysis looks at how an AC circuit performs over time. A circuit passes through a transition period before arriving at a steady-state condition, meaning a state in which the currents and voltages are not cyclical and do not vary as a function of time.

Transients can also occur in DC circuits built with energy storage elements. For instance, a sudden change in switching can trigger transient changes by assigning different conditions to circuit elements, generating sources, and the opening and closing of switches.[17]

As generators were made bigger and more powerful, transient became more of a problem. Reason: The greater voltages and power of large electrical systems meant that transient surges on their lines were potentially more destructive. Transients in switches, lights, motors, and other transmission lines produced voltage surges that spread throughout the network, often destroying line insulators, transformers, and generators.[18]

Unlike steady-state AC circuits that are stable, transient means the circuit is unstable. If the constants of an electric circuit—including resistance, inductance, capacity, disruptive strength, and voltage—have values that cannot coexist, the circuit is unstable, and remains so as long as these constants remain unchanged.

In his book *Theory and Calculation of Electric Circuits*, Steinmetz gives this somewhat technical explanation of transient voltage (text in brackets is commentary I have added; the "electric arc" he refers to is covered in greater detail in chapter 5 of this book):

> The electric arc is the most frequent and most serious cause of instability of electric circuits, and therefore should first be suspected, especially if the instability assumes the form of high-frequency disturbances or abrupt changes of current or voltage, such as is shown for instance in the oscillograms [oscillograms are the display of a wave form by an oscillograph, a device that can closely monitor rapid changes in the quantity under observation; modern versions are called oscilloscopes—see fig. 3-6].
>
> Somewhat similar effects of instability are produced by pyro-electric conductors [crystals capable of conducting electricity].
>
> Induction motors and synchronous motors may show instability of speed; dropping out of step, etc. [In synchronous motors, the shaft rotates in synch with the current frequency.]

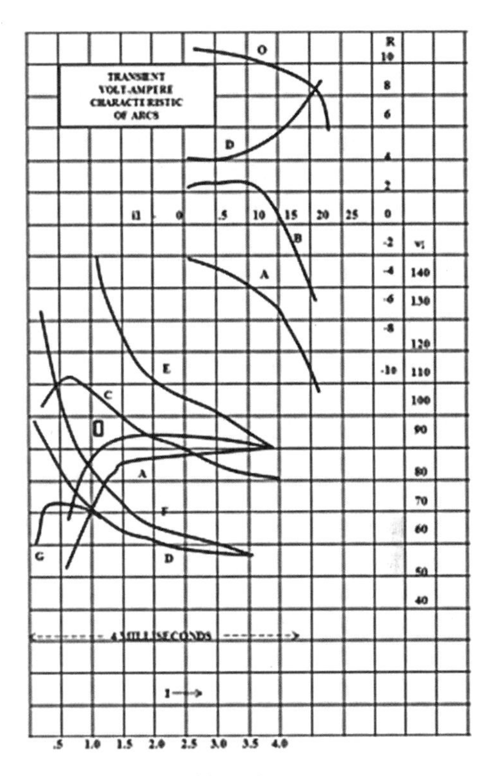

Fig. 3-6. Transient current.

The most interesting class in this group of unstable electric systems are the oscillations resulting sometimes from a change of circuit conditions (switching, change of load, etc.), which continue indefinitely with constant intensity, or which steadily increase in intensity, and may thus be called permanent and cumulative surges. They may be considered as transients in which the attenuation constant is zero or negative.

In the transient state resulting from a change of circuit conditions, the energy which represents the difference of stored energy of the circuit before and after the change of circuit condition is dissipated by the energy loss in the circuit. As energy losses always occur, the intensity of a true transient thus must always be a maximum at the beginning, and steadily decrease to zero or permanent condition.

An oscillation of constant intensity, or of increasing intensity, thus is possible only by an energy supply to the oscillating system brought about by the oscillation. If the energy supply is greater than the energy dissipation, the oscillation is cumulative and steadily increases until self-destruction of the system results, or the increasing energy loss becomes equal to the energy supply, and a stationary condition of oscillation results.

The mechanism of this energy supply to an oscillating system from a source of energy differing in frequency from that of the oscillation is still practically unknown and very little investigating work has been done to clear up the phenomenon. It is not even generally realized that the phenomenon of a permanent or cumulative line surge involves an energy supply or energy transformation of a frequency equal to that of the oscillation.[19]

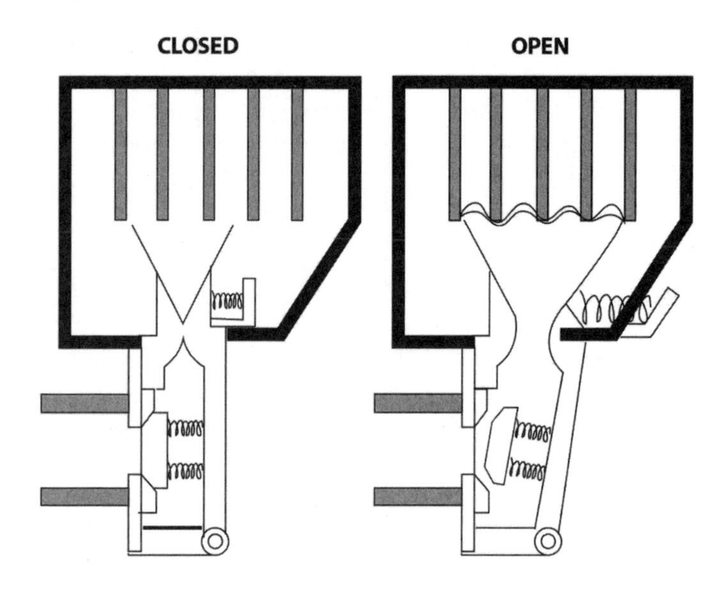

Fig. 3-7. Circuit breaker closed (left) and open (right).

High-voltage switches

Circuit breakers are protective relays found in household, commercial, and other electrical systems. When the electric current in a particular circuit (e.g., the wires going to an electric oven, central air-conditioning unit, lathe, or drill press) becomes too high, the circuit breaker automatically clicks open, "breaking" the closed circuit. We say the current overload has "tripped" the circuit breaker, putting a break or gap in the circuit. This stops the flow of current before the overload starts an electrical fire (fig. 3-7).

In the 1960s, fuses became commonly used instead of circuit breakers in many homes, including my childhood home. When the circuit overloaded, the fuse would break the circuit by burning out; to restore power, the old fuse had to be removed and discarded and a new fuse put in.

Within the fuse box was a heavy metal block with a handle, inserted into a rectangular cavity, also lined with metal. This was in essence a circuit breaker for the entire household current. It was not automated. It had to be pulled by hand, and the purpose was to enable the homeowner to manually kill the power in case of emergency.

Naturally, being a curious 11-year-old, I waited until my parents were not home, went to the basement, and pulled out the circuit breaker by its insulated handle. As I was removing the household circuit breaker from its receptacle, jagged bolts of electricity—small electrical arcs resembling miniature lightning—jumped between the metal block and metal walls of the receptacle. Then I pulled it a bit farther away, at which time the circuit was completely broken, the lightning arcs stopped, and the electricity went out throughout the house. Warning: Do NOT try this at home!

Decades before my electrical mischief, Steinmetz also knew that switching off the load from a high-voltage system resulted in an arc discharge between the contacts of the switch. Two companies competed in making circuit breakers for high-voltage systems. Westinghouse, not surprisingly, used a switch that broke the circuit in open air through slow separation of the contacts. I say "not surprisingly" since George Westinghouse's most famous invention was an air brake for trains. The oxygen-rich air, however, provided fuel for a flame the electric spark sometimes ignited and was therefore not terribly safe.

At GE, where Steinmetz was employed, the company designed a circuit breaker that extinguished the electric arc in oil, which was safely kept in a small tank lined with brick partitions. But some industrial engineers were concerned that the quick action of the GE oil switch might set up destructive power surges on the transmission lines.

To determine whether this would in fact happen, Steinmetz conducted tests on a small private grid at the GE plant. The GE pilot plant consisted of a half mile of high-voltage cable, a circuit breaker, a spark gap to measure voltages, and a 10,000-volt alternator—a generator producing alternating current. The spark gap consisted of a pair of charged metal spheres with an open space or gap between them.

While measuring surges produced by various types of open circuit breakers, Steinmetz found that when he shortened the spark gap, the result was a thirtyfold increase in electric force from 10,000 volts to 300,000 volts: the shorter gap produced a more intense electric arc. The air within this tiny gap quickly became superheated, which caused a small explosion.

Such a circuit in essence combines the steady state, which is the uninterrupted flow of current without fluctuation, with a transient circuit, in which the current values fluctuate. As we have already discussed in this chapter, Steinmetz had worked out the behaviors of both steady-state and transient voltages, so he was uniquely equipped to solve the circuit breaker problem.

To simplify his calculations, Steinmetz modeled the transmission line as having all its capacitance concentrated at the midpoint of the cable, rather than regarding it as distributed evenly along the line. The latter is more accurate, but the former is a close enough approximation and much simpler to work out mathematically.

Steinmetz calculated that the frequency of the line oscillations was a function of the qualities of the transmission cable carrying the voltage and not of the alternator generating the voltage. He found that the maximum voltage rise in the transmission line when the power was switched from off to on was less than twice the generator voltage—low enough for air to safely maintain an electric arc. However, switching on an already loaded line could produce up to a fourfold increase, and rupturing a short circuit could multiply the voltage tenfold.

Steinmetz and his GE team performed field tests of both Westinghouse air and GE oil circuit breakers in 1901 on a 44-mile, 25,000-volt line at Kalamazoo, Michigan, which had been installed just two years before. Result: as Steinmetz's calculations had predicted, the GE oil switches worked best. Just as the mathematics had indicated, low-frequency oscillations of less than twice the generator voltage were produced when the GE oil switch opened a loaded line. By comparison, when a Westinghouse air switch opened a loaded line, it produced voltages three or four times greater than the generator, triggering a self-interrupting short-circuit arc. The oscillograph, an early type of oscilloscope, showed that Steinmetz's calculations of the frequency of these oscillations were also correct: the oil switch opened the circuit at the zero point of the AC wave, making the quick action of this switch harmless.

The only discrepancy between Steinmetz's equations and equipment performance took place when *closing* open-ended power lines. In those instances, low-frequency surges (80 to 90 hertz)—up to three times the generator voltage—took place no matter what switch was used. Worse, they lasted longer than higher-frequency surges.[20]

4

The Current Wars

Though many men made significant contributions to the development of electricity, four were central to the building of the American electrical power grid as we know it today: Thomas Edison, George Westinghouse, Nikola Tesla, and Charles Steinmetz. (See page 150 for other major figures in the development of electricity.)

The stakes were high: whoever's system eventually prevailed would stand to make an enormous profit, shutting out his competitors. Actually, the outcome of the current war went way beyond which inventor would become rich or famous or both: nothing less than the future of our nation—its prosperity, growth, and progress—hinged on ensuring that the right technology was chosen for the grid that would power the country. The "war" would decide whether direct current (DC) or alternating current (AC) would become the standard of distributed electrical power in America.

Three of the men in the current wars—Steinmetz, Tesla, and Westinghouse—believed AC was the future. Edison, the most prodigious inventor of the bunch, was a staunch advocate of DC. And so the battle began. Which would rule, AC or DC?

No arguing with Edison

Edison and Westinghouse differed greatly in their approaches to electricity. Edison was dedicated to the direct current system, while Westinghouse focused on alternating current. Edison's direct current power systems had some early success. By the 1880s, he was operating about 1,500 direct current power stations nationwide. Some supplied dedicated power to a single factory or industrial campus. Others were central stations providing power to the public. But the transmission area for those central stations was limited to about a square mile, because direct current cannot travel efficiently over long distances, whereas AC can.[1] All the DC power stations were isolated, meaning they were not interconnected to one another.

Both Edison and Westinghouse were smart, wealthy, and self-made, and both had large egos. Each being the head of a company, they were used to getting their way and having subordinates follow their orders.

But Edison did not answer to Westinghouse and vice versa. Their relationship was, to degrees that varied over the years, adversarial. Their debate over DC versus AC current rapidly escalated from a technical disagreement to outright anger and animosity, fueled by their mutual stubbornness and obstinacy. You have heard the expression about two people agreeing to disagree. When it came to electrical current, Edison and Westinghouse could agree on almost nothing.

Edison went to great lengths to beat out Westinghouse, Steinmetz, and alternating current. One of his dirtier tactics was a PR campaign to convince the public that AC was dangerous and deadly. In this effort Edison was aided by Harold Brown, an electrical engineer who, like Edison, believed that AC was a hazard and threat to public safety.

In December 1888, Edison brought Brown to his facility at Menlo Park, New Jersey, for the express purpose of using Westinghouse's AC equipment to electrocute animals, to prove its deadly power. For their anti-AC smear campaign, Edison and Brown electrocuted numerous dogs, two calves, and a horse. Edison is vilified for using alternating current to execute an elephant, to even more dramatically demonstrate AC's deadly power with the largest land animal on the planet. Topsy, an elephant kept at Luna Park on Coney Island, was electrocuted in 1903 using 6,600 volts of AC power produced by General Electric generators in nearby Bay Ridge, operated by the Edison Electric Illuminating Company of Brooklyn.

Although the latter company was branded with Edison's name, he played no role in its formation or management, or in the execution of the elephant. In fact, Edison Electric carried out the electrocution of Topsy the elephant at the specific request of Luna Park officials and in complete cooperation with the Society for the Prevention of Cruelty to Animals (SPCA). The reason for the SPCA's support is that Topsy had killed three men over a three-month period and had also menaced the local police and some workmen; the elephant was clearly dangerous and a proven—and given his size, uncontrollable—man-killer. The SPCA believed then that electrocution was more humane than other methods available for putting down such an enormous animal.[2]

These cruel acts achieved their publicity goal, getting the *New York Times* to write: "The experiments proved the alternating current to be the most deadly force known to science, and that less than half the pressure used in this city for electric lighting is sufficient to cause instant death."[3] In addition, the Edison sales force, most likely following Edison's orders, began accusing Westinghouse of lying about the advantages of alternating current; it got so bad that George Westinghouse briefly considering suing.[4]

Edison would consider no alternative to his direct current system of electrical distribution. To aid him in his battle in the current wars, Edison hired another electrical genius to work for him, Nikola Tesla. But contention grew between the two men shortly thereafter.

Enter Nikola Tesla

Like Steinmetz, Tesla was an immigrant; he was born in Croatia, studied at the Austrian Polytechnic School in Graz, and came to New York in 1882. Tesla had little personal involvement with Steinmetz if any at all, but their work often intersected, overlapped, and sometimes competed.

In the early 1880s, Tesla was hired by Edison. But they fundamentally disagreed on the direction the development of electrical power should take. In addition, they were polar opposites in personality. Edison was a direct, blunt, plain-speaking man who liked plain, simple things. Tesla, by comparison, was eccentric and a bit of a dandy who craved luxury and the finer things in life. For instance, Tesla carefully and thoroughly polished with fine linen napkins his silver and crystal before he ate.

Tesla lived most of his adult life in a series of rooms in luxury hotels. Clearly suffering from obsessive compulsive disorder (OCD), he would walk around the block three times, carefully avoiding stepping on cracks in the sidewalk, before entering the hotel. His dearest companion was a pet pigeon, white with light gray on its wings, who lived in his hotel room with him. He calculated the cubic contents of each course he was served at dinner before he would eat a dish. He required a fresh tablecloth with every meal, two dozen napkins, and silverware that had been sterilized. Many commonplace things he found intolerable, such as women wearing earrings; Tesla never had a romantic relationship with any women.[5]

Edison had promised Tesla a $50,000 cash bonus—an enormous sum then—if Tesla could significantly improve the performance of Edison's dynamo, a generator producing direct current. Tesla designed the improvements. But then Edison welched on the promised bonus, saying Tesla had misunderstood; Edison had just been joking about a bonus. After all, why should Edison pay a bonus to Tesla, whom he viewed as simply another (albeit brilliant) engineer on the Edison payroll, doing the job he was already being paid to do? Angered at being cheated, Tesla quit his job with the Edison Company and started a rival firm, which would focus on developing AC as a superior alternative to Edison's DC. Tesla was convinced that alternating current was superior for long-distance power transmission and that Edison's direct current was the wrong choice. Tesla held dozens of patents on his AC induction motor and polyphase AC system.

In his autobiography, Tesla said that in 1893 he built a conical coil that generated 100 million volts, which he figures was equivalent to a flash of lightning. He built a transformer that produced current "many times stronger than in the usual way" and created a spark over 100 feet long. Tesla added, "Judging from my past experience there is no limit to the possible voltage developed; any amount is practicable."[6]

In his book *Experiments with Alternate Currents of High Potential and High Frequency*, Tesla expressed his delight at exploring the mysteries and promise of AC:

> One reason, perhaps, why this branch of science is being so rapidly developed is to be found in the interest which is attached to its experimental study. We wind a simple ring of iron with coils; we establish the connections to the generator, and with wonder and delight we note the effects of strange forces which we bring into play, which allow us to transform, to transmit and direct energy at will.
>
> We arrange the circuits properly, and we see the mass of iron and wires behave as though it were endowed with life, spinning a heavy armature, through invisible connections, with great speed and power—with the energy possibly conveyed from a giant distance. We observe how the energy of an alternating current traversing the wire manifests itself—not so much in the wire as in the surrounding space—in the most surprising manner, taking the forms of heat, light, mechanical energy, and, most surprising of all, even chemical affinity.
>
> All these observations fascinate us, and fill us with an intense desire to know more about the nature of these phenomena. Each day we go to work in the hope of discovering—in the hope that someone, no matter who, may find a solution of one of the pending great problems—and each succeeding day we return to our task with renewed ardor; and even if we are unsuccessful, our work has not been in vain, for in these strivings, in these efforts, we have found hours of untold pleasure, and we have directed our energies to the benefit of mankind.[7]

In 1886, George Westinghouse bought many of Tesla's most significant patents, because he, like Tesla and Steinmetz, believed the future of electricity was AC, and he intended to profit handsomely from it. Tesla was always in need of funding because, unlike Edison, Westinghouse, and Steinmetz, who were either rich or worked for well-capitalized employers, he did not have the resources of a big corporation to back him up.

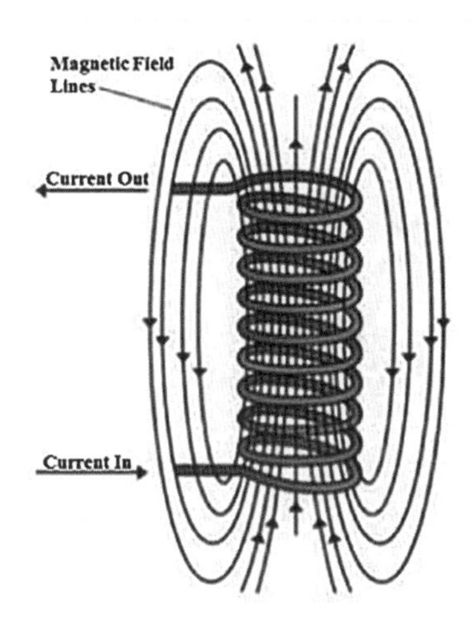

Fig. 4-1. Tesla coil.

Tesla suffered frequent business reversals. Once, his company's lab with most of his apparatus and experiments burned to the ground. Also, several times he was evicted from the expensive hotels in which he lived as a long-term resident, and left owing huge sums on his unpaid hotel bills.

Despite his lack of cash, Tesla, like Edison, was a prolific inventor. Variations on his original Tesla coil (fig. 4-1) are used today in radio and television receivers. Tesla introduced to the world the fundamentals of robotics, computers, missiles, microwaves, nuclear fusion, and satellites. He even designed a beamed weapon that he called a "death ray" (fig. 4-2).

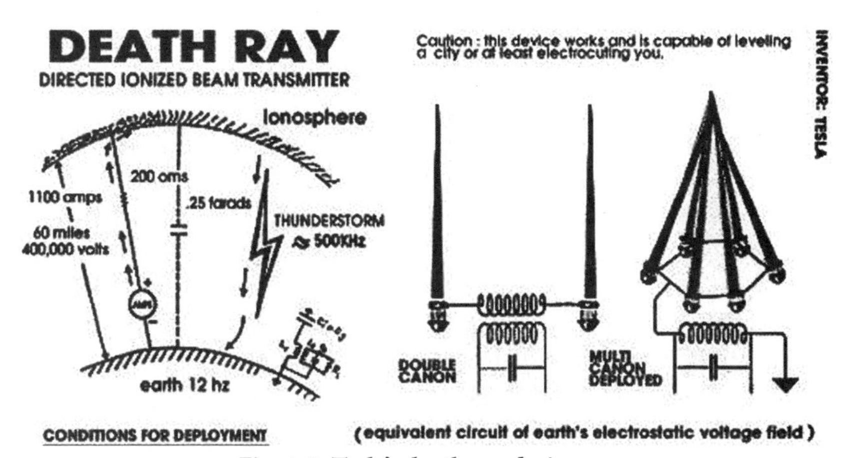

Fig. 4-2. Tesla's death ray design.

George Westinghouse.

Westinghouse: the man with the money

Like Tesla, Edison, and Steinmetz, George Westinghouse was an engineer, inventor, and industrialist. But unlike Edison and Tesla, who acquired wealth to fund their science, Westinghouse desired the money to make himself rich and powerful for its own end.[8] Steinmetz took a different path; he was indifferent to wealth and was content to work for a corporation. As long as he had the freedom and resources to pursue his research, he was happy.

Of the four men, Westinghouse clearly had the most business acumen and built the biggest net worth: GE was on a par with Westinghouse in terms of size, but while George Westinghouse was the boss at Westinghouse, Steinmetz was strictly an employee, and not an owner or chief executive at GE. Today, Westinghouse has annual sales of nearly $5 billion and GE annual sales of about $20 billion.

With the combination of technical ability and business skill, Westinghouse was uniquely positioned to put his prototype inventions into production and sell them for massive revenues. Of the AC versus DC current war, Tesla commented, "George Westinghouse was, in my opinion, the only man on the globe who could take my alternating current system under the circumstances then existing and win the battle."[9]

As an inventor, Westinghouse is best known for designing the railroad air brake (fig. 4-4). Introduced in 1869, the Westinghouse brake applied compressed air to produce pressure much greater than human muscle, providing superior stopping power for trains with greater safety. The industrialist also invented signaling equipment for railroads.

Plus, Westinghouse held patents on natural gas piping systems and equipment. His system transported natural gas through the pipes under controlled conditions over long distances. With the Westinghouse system,

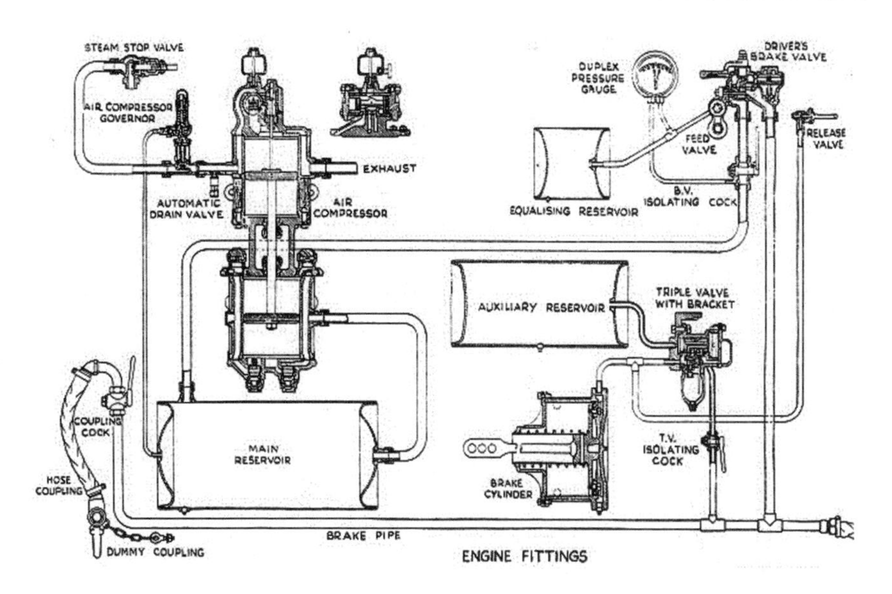

Fig. 4-4. Westinghouse train air brake.

gas could be piped to homes and apartment buildings for fueling ovens and furnaces with natural gas.

In addition to the natural gas patents, Westinghouse bought numerous electricity patents from Tesla. He was a patron of sorts for the Croatian scientist: when Tesla could not afford to pay his rent, Westinghouse paid for Tesla's room and board at the Waldorf Astoria Hotel in New York City. While Westinghouse was rich, Tesla struggled financially; he was a genius scientist but, unlike Edison and Westinghouse, a lousy businessman who died in near poverty. In fact, after George Westinghouse passed away, the executives at Westinghouse apparently felt some guilt, knowing that their founder grossly underpaid Tesla for the patents, given the enormous value of these innovations to the Westinghouse Company. To make up for it, they offered to pay Tesla's rent for the rest of his life.

By applying Tesla's innovations and (for a time) enlisting Tesla's aid, Westinghouse is credited with being the first to commercialize AC generation and transmission systems.

Proving the case for AC

To gain the support of both the public and the government for either AC or DC, the demonstration of the electrical distribution system had to be on a grand scale.

In 1893, the U.S. government held a competition to see which electrical firm would be awarded the contract to provide lighting for the Chicago World's Fair, which would be open to the public for six months. The competitors were General Electric, Westinghouse Electric, and Edison Electric, the first two offering an alternating current system and the latter direct current. As is often the case with government projects, the contract was awarded to the low-price bidder, which was Westinghouse. Remember, Westinghouse owned Tesla's patents, which the company used to design a polyphase alternating current system (we'll explore polyphase systems a bit later in this chapter).

The Westinghouse AC distribution system was proven successful in spectacular fashion. The Westinghouse design powered 92,000 outdoor incandescent lamps and 250,000 interior lamps. Because Edison lost the bid, he responded to the rejection by refusing to allow his light bulbs to be used. (The World's Fair was not a total bust for Edison. He readily took advantage of the fair to promote several of his other inventions, which were demonstrated in the fair's showcase electrical building.)

To solve the problem, Westinghouse designed and perfected an alternate light bulb called the Sawyer-Man stopper lamp. Unlike Edison's one-piece light bulb, the Sawyer-Man had two pieces: (a) a low-resistance filament inserted into (b) an iron-and-glass "stopper." The stopper was fitted into a glass globe filled with nitrogen and then sealed. In an Edison bulb, once the filament burned out, the bulb had to be replaced. In the two-piece Sawyer-Man, the stopper could be removed from the bulb and the filament replaced, making it the first reusable light bulb.[10]

Why did Westinghouse keep dropping his bid price until he wrested the contract away from Edison and his other competitors? Companies generally lowball their price for two reasons. The most common reason for underpricing is what businesspeople call "buying the business." This means you price your product or service low deliberately to get your foot in the door with a new customer, with the hopes of getting more orders, even though you may be losing money on the initial order. The second reason, which turned out to be of incalculable value to Westinghouse, was the free publicity from being the one to light up the World's Fair. He could not have bought this much advertising for $100,000!

In business, a common venue for bringing a new product or idea to the attention of the public is to hold a press conference. Reporters are invited to a lecture and demonstration of the new technology, often conducted by the company president along with the chief technology officer.

However, Westinghouse unknowingly took advantage of a more modern PR technique, described by New York publicity expert Eric Yaverbaum in his book *Public Relations for Dummies*. Yaverbaum advises businesses not to hold press conferences, which are often poorly attended. Instead, he tells publicists to "go where the public and press already are," which gives the publicist a ready-made audience. And there was no bigger stage at the time than the World's Fair, which was teeming with press. Plus, demonstration of the alternating current system as the official lighting producer of the fair gave Westinghouse an unbeatable endorsement.

Westinghouse's tremendous success at the Chicago World's Fair led to an even bigger and more prestigious contract for Tesla's system: the power plant at Niagara Falls. The Niagara River flows north from Lake Erie to Lake Ontario, dropping roughly 188 feet, making it the largest waterfall in the eastern United States. (The second biggest is the Great Falls of Paterson, New Jersey, where Alexander Hamilton build a textile mill to take advantage of the power generated by the falls.) More than a hundred years before the installation of an electrical power plant there, early settlers were already using Niagara Falls for power. They dug a short canal loop near the base where the water falls. The canal had its intake near the upstream end of the rapids and its discharge above the falls, and a waterwheel on the canal powered a mill.

Building a power generation system at Niagara Falls had been a dream of Tesla's ever since he was in high school, when he envisioned a huge waterwheel being turned by the powerful falls. He told an uncle he would go to America to carry out this plan.[11] Once again, GE and Westinghouse were in direct competition to build an electrical power plant and distribution grid. Only this time, George Westinghouse became suspicious that industrial spies from General Electric were stealing his plans for the Niagara Falls power system. In point of fact, a Westinghouse draftsman was arrested for secretly selling the firm's World's Fair and Niagara Falls blueprints for thousands of dollars to two GE employees. When a search warrant was issued, the plans were found at GE's offices. But when the case went to trial, the jury was deadlocked.[12]

However, a new firm, Cataract Construction Company, won the Niagara Falls contract. It built a power plant based on a dozen Tesla patents. The new plant had three polyphase generators of 5,000 horsepower each; combined they produced 2,000 volts. Transformers, based on the Tesla Coil design, multiplied the current fivefold to 10,000 volts. Copper conductors on wooden poles with porcelain insulators carried the current 26 miles to Buffalo, New York. In Buffalo, a substation used transformers to step the current back down to 2,000 volts.

To drive the turbines with rushing water, a large tunnel was dug underground for the hydraulic transmission system. The horseshoe-shaped tunnel was 21 feet high, 19 feet across at its widest point, and 6,700 feet long. A crew of as many as 2,500 men at a time took over two years to dig and build the structure. They excavated 600,000 tons of rock and earth and reinforced the tunnel using 16 million bricks, 19 million feet of timber, and 67,000 barrels of cement.[13]

The success of both the World's Fair and the Niagara Falls projects put AC firmly in the lead of the current wars versus DC. But Steinmetz wasn't done yet. In 1907, Steinmetz built a transformer capable of raising the power of electricity to 220,000 volts, which was the highest load ever carried up to that time. He received a patent on the improved high-tension cable he designed that was capable of carrying that amount of power. In this he was truly ahead of his time, because back then the greatest amount of power any factory needed was 80,000 volts.

Edison's bid for direct current was destined to fail right from the very start; AC had too many inherent advantages over DC. Alternating current is more powerful than direct current. It is also more economical. Unlike direct current, which required a power generator every mile and copper conduits buried under the streets to carry the current, alternating current could travel for hundreds of miles, either below or above ground, without losing strength or requiring intermediate generators or transformers along the transmission circuit to boost the power.

And whereas the voltage of direct current cannot be changed, transformers can either increase or decrease the voltage of alternating current. This ability to step up or step down the current's voltage enables the grid to carry electricity over hundreds of miles of cable. Also, in an AC system, wires can carry a thousand times more electricity than in a DC grid.[14]

In his book *Steinmetz: Engineer and Socialist*, Ronald Kline neatly summarizes why DC was doomed from the very start to lose out in the current wars to AC:

> Electrical manufacturers and utility companies turned to alternating current because of the limitations of Edison's direct-current system. AC varies in direction and intensity at a prescribed frequency, 60 times per second for household current in the United States. DC flows in one direction and is steady, like the current from a battery.
>
> The DC system was practical only in densely populated urban centers, because the greater the distance from the dynamo to the customer, the larger in diameter the distribution wires had to be to make the system economical (energy losses are inversely proportional to the diameter of the wire). Engineers first attacked the problem by raising the voltage of the system with a three-wire 220-volt feeder network, which increased the service area to only two miles or so.
>
> AC proved to be a better solution because of its flexibility. It could be generated at a low voltage, stepped up to a high voltage in the thousands of volts for distribution, then stepped down to a low voltage for home use—the system in general use today.[15]

The greatest value Steinmetz brought to the table in the victory of AC over DC was his rigorous and accurate mathematical analysis of the behavior of alternating current. The mathematics of AC are much more complex than DC. Direct current is relatively simple: it goes on one direction and can be precisely measured with a meter.

By comparison, alternating current has no definite direction. Its direction and values—both voltage and amps—are constantly changing, which makes measuring its flow, predicting its behavior, and designing equipment for efficient transmission of AC a much more complex task than DC. And it is a mathematical complexity that Steinmetz mastered as no other man on Earth had ever done.[16]

Losing both the Niagara Falls and the World's Fair projects to AC systems based on Tesla's designs and Steinmetz's theories of electrical current essentially put Edison and his direct current out of that business, and the victory in the electricity war was won by AC.[17] The big World's Fair contract, unfortunately, did not make Westinghouse immune to the ups and downs of life and business. He lost most of his fortune in the Panic of 1907. But the Westinghouse Company survived and flourished; I was an employee there in the late 1970s. In 1955, George Westinghouse was elected to the Hall of Fame for Great Americans.

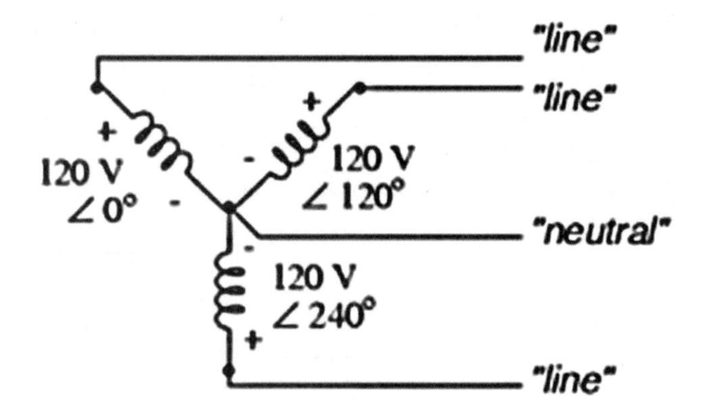

Fig. 4-5. Three-phase power plant circuit design.

The polyphase AC system

Steinmetz made breakthroughs in both DC and AC but is largely credited with making today's modern alternative current power grid reliable, efficient, and able to transmit power over long distances.

There are two reasons why it is fortunate that the alternating current technology developed by Tesla, Westinghouse, and Steinmetz beat out Edison's direct current for America's electrical power distribution grid. The first reason is that it is relatively easy to convert alternating current to direct current, which is in fact what Steinmetz did when developing his electric arc street lamps (see chapter 5). On the other hand, converting DC to AC is difficult and expensive. The second reason AC was the more sensible choice is that, thanks in large part to the pioneering work of both Tesla and Steinmetz, AC can be transmitted over much greater distances over the grid.

Most large power plants produce what is known as "polyphase" alternating current, meaning the current is generated and transmitted in multiple phases simultaneously.[18] AC power systems can have one, two, three, or four phases. In a two-phase system, for instance, a device is powered by two independent, alternating currents that are said to be 90 degrees "out of phase," meaning that the second current begins when the first one has peaked.

Many polyphase power plants produce three different phases of AC power rather than just one or two. Three-phase power plants produce large amounts of electricity more efficiently than single-phase current (fig. 4-5).

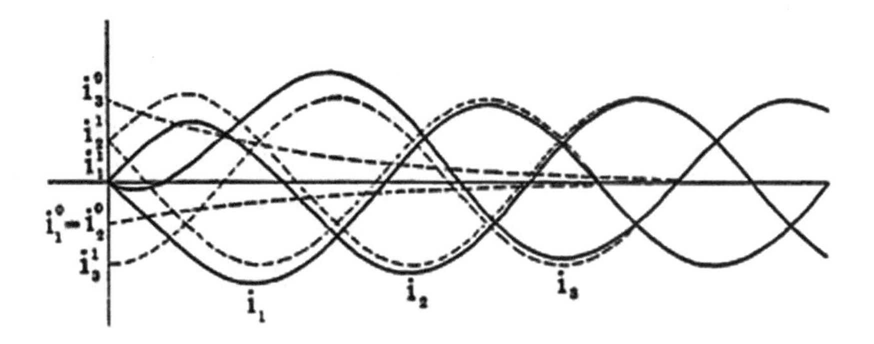

Fig. 4-6. Current in three-phase electrical system.

The three phases are synchronized and separated by 120 degrees from one another. There are four wires coming out of every three-phase power plant. Each phase has one wire, and there is a grounding wire shared by all three.

Alternating current is propagated as a sine wave. In a one- or two-phase system, the sine wave crosses zero volts 120 times a second. In industry standard three-phase power, at any given moment one of the three phases is nearing a peak (fig. 4-6).[19] As a result, three-phase motors and other equipment always have the electricity they need to operate without interruption. Adding a fourth phase would require another wire, but the power output would not improve significantly. Therefore, three-phase has become the standard in polyphase AC systems. The world's first polyphase system, consisting of dual three-phase generators rated at 250 kilowatts and 2,400 volts, was built by General Electric in Redlands, California, in 1895.[20]

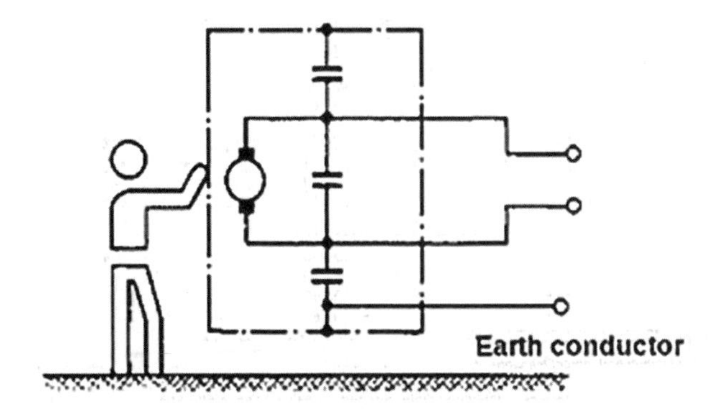

Fig. 4-7. Electrical circuits are grounded when connected to the ground.

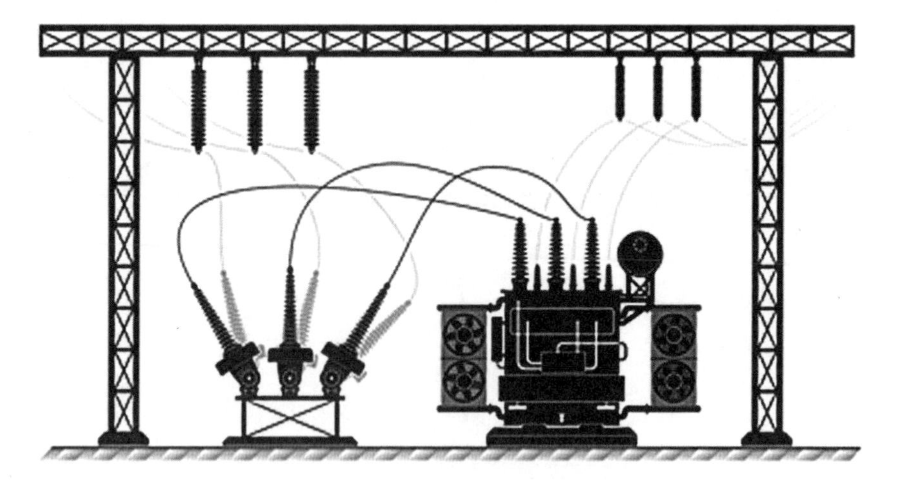

Fig. 4-8. An electrical substation.

As mentioned above, in the three-phase design there are three wires, one for each phase, and a fourth wire called the "ground" wire. It is so named because it connects with the ground (fig. 4-7). Why have a wire connect to the ground? Well, the earth is a good conductor, and there is no bigger conductor for grounding available than the actual planet we live on. The earth provides a return path for electrons. "Ground" in the electrical grid is the actual ground we walk on.

Why is grounding desirable in electrical circuits? The main reason is that it protects your appliances, your home, and you from electrical surges, which could be caused by instability in the utility grid distribution system or a lightning strike. If your electrical system is grounded, all that excess electricity goes directly into the earth, without damaging your appliances or burning out your wiring system.[21] It can also protect you from harm or even electrocution. The National Electric Code gives detailed grounding specifications for building wiring to protect structures against fires started by electrical overloads.

Towers of power

When driving, you occasionally come across extremely tall metal towers with cables strung from tower to tower. These cables are the utility grid's high-voltage transmission lines. In the grid, three-phase power leaves the generator and enters a transmission substation (fig. 4-8), where it passes through a transformer. The voltage produced by the generator is a few thousand volts. The main transformers at the substation step up the current to 155,000 to 765,000 volts. The voltage has to be this high to prevent energy loss over long-distance transmission lines. The typical maximum transmission distance between substations, which are mainly clusters of transformers, is about 300 miles.

Tesla is credited with developing the first transformers, using his Tesla coil as the basic circuit design. These transformers could raise or lower the voltages. Steinmetz's first American employer, E&O, was known for manufacturing quality transformers.

Voltage is simply a measure of the force with which electricity flows through a wire. The higher the voltage, the more power transmitted. When the current arrives at the power substation nearest your home, a medium-size transformer reduces the voltage. Then as it flows through local power lines from the substation to your house, smaller transformers on the poles or underground further reduce the voltage. The electricity generated by a power plant's generators is at such a high voltage that it would destroy your home if it was not reduced by transformers first.

5

A Street Lamp Named Steinmetz

Poet Dylan Thomas famously wrote, "Rage, rage against the dying of the light."

And throughout all of human history, we have done just that. Early humankind feared the dark. Our ancestors discovered fire, produced by the combustion of wood, and used it to light their caves and camps at night to beat back the dark. To this day, many people are afraid of the dark. Psychologists call extreme fear of the dark night phobia, or nyctophobia. Symptoms it can cause in both children and adults include rapid shallow breathing, heart palpitations, shivering, trembling, nausea, crying, and screaming.[1] No wonder the world so welcomed the light brought to it by Steinmetz, Edison, and their fellow electrical pioneers!

Long before Edison invented the light bulb and Steinmetz and others built a system for reliable distribution of electric power, we kept our homes illuminated with fireplaces, candles, and kerosene lamps.

But what about away from home, walking about town or riding a horse and buggy? Answer: street lamps. And these illuminated glass globes and containers gave light long before electricity was widely used for power or could be transmitted more than extremely short distances.

In 1667, Paris became the first city in the world to install street lamps, which consisted of wax candles in glass containers.[2] In London, which also used street lights before the United States, the lamps consisted of a wick burning in oil contained in a glass globe. The smoke darkened the interior of the globes, which necessitated frequent cleaning.

It was Ben Franklin, best known in science for capturing the electrical power of a thunderstorm with a kite and key, who designed the first street

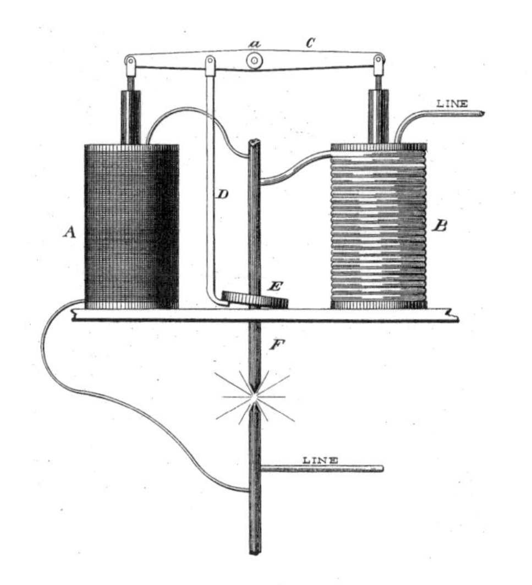

Fig. 5-1. Brush's arc lamp.

lighting in America and convinced the Pennsylvania Assembly to install these lamps in 1757. Franklin's innovation was to encase the oil candles in a housing of four flat panes of glass. Crevices at the bottom allowed entry of air, and a long funnel at the top drew up the smoke and exhausted it from the top of the tube, preventing the glass from blackening.

In 1803, Newport, Rhode Island, installed street lamps illuminated with gaslights. In 1880, Thomas Brush, a competitor of Edison's, built the first electric streetlights, called brush lamps (fig. 5-1), in Wabash, Indiana. The Brush lamp was widely used in streetlights and in commercial and public buildings. The lamp combined an arc light and a dynamo. The dynamo generated DC power for the lamp, resulting in high-intensity illumination.[3]

As towns and cities across the country desired to light up their own streets, a number of engineers, including Steinmetz, began investigating whether they could improve on Brush's design.[4]

In his book *Radiation, Light, and Illumination*, Steinmetz goes into great detail explaining the lighting requirements of streets and the approach he took to solving the problem:

> The problem of street illumination is to produce a uniform low intensity. For reasons of economy, the intensity must be low, at least in American cities, in which the mileage of streets, for the same population, usually is many times greater than in European cities, and, at the same time, the same type of illuminant is usually required for the entire area of the city. The low intensity of illumination requires the quality of light which has the highest physiological effect at low densities, that is, white light, and excludes the yellow light as physiologically inefficient for low intensities. Still better would be the bluish green of the mercury lamp, but is not much liked, due to its color. Quite satisfactory also is the greenish yellow of the Welsbach mantel for these low intensities. The American practice of preferring the white light of the carbon or magnetite arc thus is correct and in agreement with the principles of illumina-

Fig. 5-2. An early street lamp.

> tion, and the yellow-flame arc can come into consideration even if it were not handicapped by the necessity of frequent trimming only in those specific cases where a high intensity of illumination is used, as would be only in the centers of some large cities.
>
> Uniformity of illumination is especially important in street lighting, where the observer moves along the street, and, due to the low intensity, the decrease of subjective illumination by fatigue is especially objectionable. For a street illuminant, a distribution curve is required which gives a maximum intensity somewhat below the horizontal, no light in the upper hemisphere, and very little downward light. Street lamps therefore should be judged and compared by the illumination given midways between adjacent lamps, or at the point of minimum intensity, or, in other words, by the intensity in a direction approximately 10 degrees below the horizontal. This also is in agreement with American practice.
>
> However, it is very important that the downward intensity be very low, and in this respect it is not always realized that the light thrown downward is not merely a waste of light flux, but is harmful in producing a glaring spot at or near the lamp and, by the fatigue caused by it, reducing the effective illumination at the minimum point between the lamps. Most objectionable in this respect is the open direct current carbon arc and those types of lamps giving a downward distribution, but even with the enclosed arc lamp the distribution of light on the street surface is still far from uniform, and the intensity too high near the lamp, and in this respect improvements are desirable.[5]

Take note of the level of precision and detail in his description of the ideal streetlight. It may seem obsessive or overkill to you. But if you are a scientist, engineer, or programmer, you know that a complete and accurate definition of a technical problem greatly increases both your odds of solving it and the quality of your solution. In fact, IBM for many years offered a training workshop on "Defining Requirements" to both the technical professionals who designed systems and the users who asked for and used them.

Outlining clearly and in nontechnical terms the optimal lighting conditions for street lamps—and making the case that a magnetite arc would ideally fulfill those requirements—were key to getting GE management's buy-in and support on Steinmetz's magnetite-based design. In the corporate world, the most common complaint among system users and corporate management is that technical professionals deliver systems that are overbudget and late and do not meet the requirements asked for. In response, the technical department complains that users and management don't understand what they are asking for, don't know what they really want, can't articulate what they want, and do not understand what is possible within the limited time and budget provided.

Steinmetz avoided this dilemma, partly with his problem-solving genius, but also partly with his exceptional skill at analyzing, assessing, understanding, and describing system requirements. His magnetite street lamp is a classic example of how he met a complex technical challenge.

A light shines in Schenectady

For purely mathematical problem-solving, Steinmetz usually worked alone, bent over at a table with a pad and pencil. But in laboratory experiments involving mechanical or electrical apparatus, he was often assisted by laboratory assistants, or "lab boys" as they were called at GE back then. By having the lab boys do routine tasks, such as soldering or wiring or maintaining equipment, Steinmetz freed himself for maximum productivity on the tough mental work that only he could do.

But one of his lab boys, Joseph LeRoy Hayden, who like Steinmetz was an engineer, stood out from the rest. He was brighter, harder working, and affable, and he shared Steinmetz's fascination with all things electric. Steinmetz made Hayden his principal assistant on his research project to develop a new kind of streetlight, the "magnetite arc lamp" mentioned above, which radiated a bright bluish light.

Magnetite is a hard, black mineral composed of 72.4 percent iron. Steinmetz found that magnetite rods in arc lamps lasted many times longer than the conventional carbon rods most commonly used in arc lamps. In a magnetite arc lamp, the terminals are not hot enough to give light by incandescence. Instead, brightness is provided by the electric current that is carried across the space separating the lamp's electrodes. Specifically, the current is carried by a stream of "electrode vapor" emanating from the negative magnetite terminal. The electrode vapor is an emission of electrons at a velocity of several thousand feet per second. The arc providing the illumination occurs when current is conducted through the vapor between the two terminals.

By comparison, carbon rods get much hotter, so they give illumination by incandescence, meaning they glow from the heat. The high temperature they maintain makes them burn out much faster than magnetite terminals, which operate at lower temperature.

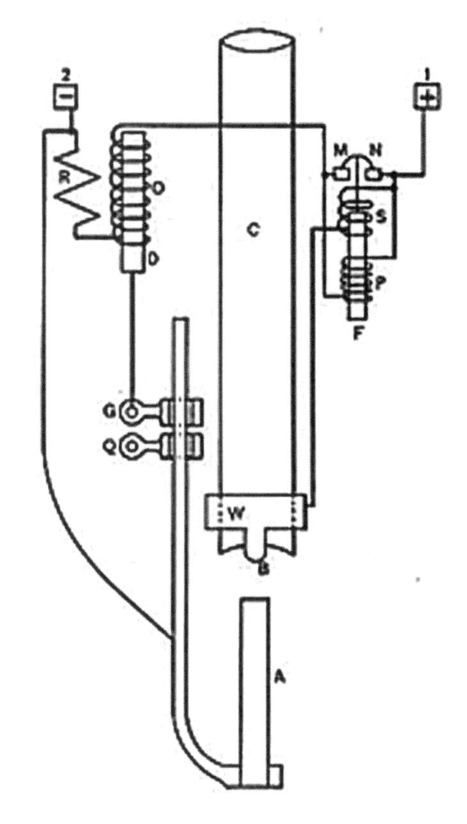

Fig. 5-3. Diagram of magnetite arc lamp with negative magnetite electrode at left.[7]

In the Steinmetz design (fig. 5-3), the magnetite lamp had an upper positive electrode of copper and a lower negative electrode of magnetite, both enclosed in iron tubes. The magnetite was mixed with titanium oxide to brighten the illumination; chromium oxide was added to extend the electrode's life.[6]

Because Steinmetz was a staunch advocate of alternating current, his new house on Wendell Avenue was naturally wired for AC. Had it been otherwise, Steinmetz would have lost huge credibility in his war with Edison, who championed DC. Imagine the publicity for Edison if his DC current was used by none other than the AC genius Steinmetz at GE!

However, the magnetite arc lamp ran on direct current. Therefore, Steinmetz had to convert the AC in the house circuits to DC for the magnetite arc lamps to operate.

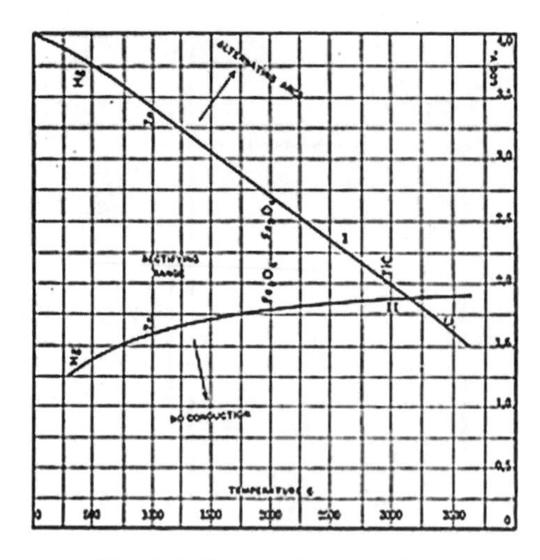

Fig. 5-4. Voltage in a rectifier.

To produce the needed direct current, he used rectifiers. These are devices in which the circuits are configured so that current flows more readily in one direction than the other (fig. 5-4). In alternating current at 60 hertz, the direction of the electron flow rapidly changes back and forth 60 times a second. By forcing the current to flow in a single direction only, the rectifier converts alternating current to direct current; and the latter flows only in one direction.

At the time of the development of the arc lamp, rectifiers did not always operate reliably. One of Hayden's responsibilities was to service and maintain the rectifiers so that the lamps had continuous DC power. Soon, in addition to rectifier repair, Hayden was given more and more responsibility by Steinmetz, and he became a valued and trusted assistant on many aspects of the arc lamp project.

During the day, Steinmetz worked in his office at the General Electric plant, where he performed his mathematical, engineering, and consultation work. But during this period, Steinmetz was having a large private home built for him on Wendell Avenue in proximity to the plant.

The house was on an odd lot that no one else seemed to want. The land sloped steeply on one side, with a swampy bog at the bottom. Steinmetz had a real estate agent show it to him, who admitted that the lot would be difficult to sell (to most people) because of these undesirable features. Steinmetz said he might be interested in purchasing the lot for construction of a new home. But, of course, because of the hilly slope and swampy section, he told the agent that he naturally expected a good price as a concession—and he got it.

What the real estate agent did not know was that Steinmetz saw the odd lot features as a positive, not a negative. He bought the land and built the house, and then turned the swampy area into a beautiful pond, about 10

Steinmetz in his home laboratory.

feet square and 5 feet deep, and stocked it full of goldfish, turtles, water lilies, swamp grasses, and other marshland plants. He built a wooden bridge across the pond. He loved to stand on the bridge in the middle of his peaceful glen, stare into the water while working on a problem in his mind, and watch the fish swim and the turtles play. On the slope uphill from the pond, Steinmetz put in a natural rock garden. Heaven on Earth—and all his![8]

One of the first parts of the home to be finished was a large, well-equipped home laboratory; the home lab was funded and equipped by GE.[9] It was a long extension off the main dwelling. The lab extension was two stories (the main house was to be three stories). On the first floor was the enormous lab, filled with workbenches, flasks, beakers, chemicals, and all sorts of tools, wire, cable, and electrical apparatus. There were two small bedrooms on the second floor.

At GE, Hayden was in charge of the power plant adjoining the laboratory. The Steinmetz home was a stone's throw from the plant, so Steinmetz and Hayden worked together on the arc light mostly after regular business hours. After working at the plant all day, both Steinmetz and Hayden would go to the laboratory wing of the new Steinmetz home, the main part of which was still under construction. There, they would work together in the laboratory, which had been completed, late into the night on the development of the arc lamp.

Hayden commuted to the GE plant from a rented room in downtown Schenectady, so Steinmetz suggested that Joe move into the new house, which would save him money on rent and eliminate his commute. Although the home was still under construction, the sleeping quarters above the lab had been completed. The two men moved into the lab, sleeping in the bedrooms above it. Steinmetz, who had no housekeeper, cooked their meals on a little gas stove. The staples of their diet were eggs, steak, and potatoes.

While Steinmetz gave the outward impression of a loner—certainly his unusual appearance and diminutive stature set him apart from other men, as did his brilliance—he in fact greatly desired the warmth and camaraderie of friends and family. Hayden's constant presence met that need, and Steinmetz was well aware of this.

Because they lived above the laboratory where they spent so much of their time conducting experiments, the relationship between Steinmetz and Hayden evolved from boss and employee to friendship. Steinmetz developed a fatherly affection for Hayden, and Hayden in return saw Steinmetz as a father figure, mentor, teacher, and friend. Joe never complained about Steinmetz's constant smoking or the ash he dropped all over the lab, although the young lab assistant eventually tired of eating nothing but steak and potatoes.

Out of the lab of this unlikely "odd couple" came the magnetite arc electrode and its bright blue-white illumination. Here is a description of Steinmetz's electric arc light from his patent for the device (fig. 5-6):

> No. 914,891 Specification of Letters Patent Patented March 9, 1909
>
> To all whom it may concern:
>
> Be it known that I, Charles P. Steinmetz, a citizen of the United States, residing at Schenectady, in the county of Schenectady, State of New York, have invented certain new and useful Improvements in Electrodes for Arc-Lights, of which the following is a specification.
>
> My present invention relates to improvements in electric arc lights and more especially to the composition of the terminals or electrodes used therein. When the terminals or electrodes are of carbon, as is now the common practice, the arc itself is of comparatively feeble intensity, the greater portion of the light being derived from the incandescent crater of the positive terminal. I have discovered however that if either or both of the terminals be made either

in whole or in part of magnetite (Fe_3O_4) instead of carbon, the arc assumes an entirely different character. Instead of being the minor factor in the production of light it becomes intensely luminous, giving off a brilliant white light with a spectrum like that of iron. If both electrodes or terminals are of magnetite, it is immaterial which is positive and which is negative. If, however, one terminal be of carbon or of some substance other than magnetite, it is desirable that the magnetite terminal should be the negative. If, with one electrode of carbon and the other of magnetite, the polarity be reversed so that the magnetite is the positive terminal and the carbon the negative terminal, the arc instead of being of the luminous character mentioned, becomes dull and comparatively non-luminous and yellowish in color, the spectrum under these circumstances presenting bright sodium and potassium lines, showing the presence of these substances as impurities.

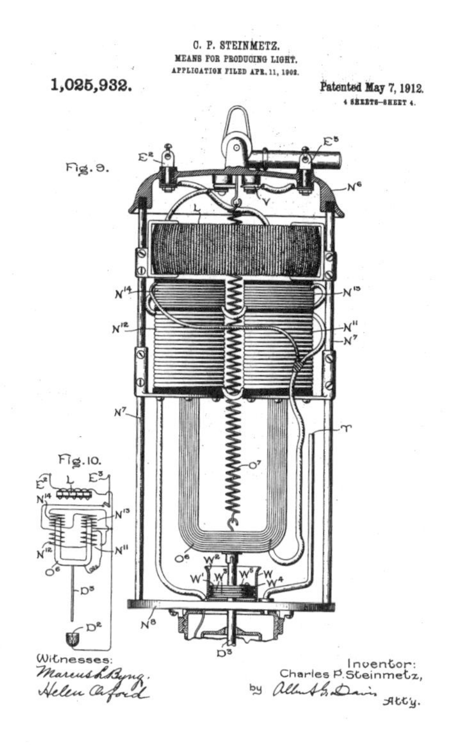

Fig. 5-6. Patent drawing of the Steinmetz electric arc.

The use of magnetite for the purposes above described not only very greatly improves the quantity and character of the light, but possesses in addition the valuable feature that electrodes formed thereof waste away only very slowly, considerably slower in fact than the wasting away of corresponding carbon electrodes. I find also that I may improve the action of the electrode by adding certain impurities such, for example, as magnesia, lime, or alumina compounds or the like. In making the electrodes, a suitable binding material may be used to render easier the formation of the electrodes out of its constituent material or materials.

In the accompanying drawing I have shown an electrode of the character in which my invention maybe embodied.

What I claim as new and desire to secure by Letters Patent of the United States, is:

1. An arc lamp electrode in which magnetite predominates.
2. An arc lamp electrode composed of magnetite.
3. An arc lamp electrode composed predominantly of a good conducting metallic oxide.
4. An arc lamp electrode composed predominantly of a metallic oxide of good conductivity and which when heated is but slightly attacked by air.
5. An arc lamp electrode which is a good conductor at ordinary temperatures, and contains a good conducting metallic oxide giving a luminous or flaming arc.
6. An arcing electrode composed essentially of an electrically conducting oxide.
7. An arcing electrode composed essentially of an oxide of iron.
8. An arcing electrode composed essentially of the magnetic oxide of iron.
9. An essentially metallic arcing electrode containing oxide of iron.
10. An essentially metallic arcing electrode containing magnetic oxide of iron.
11. An arc light electrode composed principally of magnetite.
12. An arc light electrode containing at least 75 per cent of magnetite.

In witness whereof, I have hereunto set my hand this 24th day of February, 1902. Charles P. Steinmetz.

As mentioned earlier, magnetite arc lamps gave far brighter illumination than arc lamps with carbon electrodes, and the magnetite electrodes lasted many times longer than carbon, which burned out quickly and had to be replaced often. Decades later, in 1966, General Electric introduced high-pressure sodium streetlights. Sodium gradually replaced carbon and magnetite as the standard street lamp, because it was extremely energy efficient and gave off a bright yellow glow that the public seemed to prefer.

Today, Ben Franklin's former hometown, Philadelphia, has been experimenting with using light-emitting diodes for street lamps. In a small village in France, a one-kilometer stretch of road has been paved with 2,880 solar panels, which combined produce enough energy to light every street in the town.[10]

Good housekeeping in the Steinmetz abode

After the successful development of the arc lamp, there was no talk of Hayden moving out. Both the great scientist and his lab assistant had developed a close and harmonious working relationship that ran as smoothly as a Swiss watch, and having Hayden in his home gave Steinmetz the companionship he had long hoped for.

Because of mid-project changes in the construction plans, the house took well over a year to finish. During that time, Steinmetz and Hayden continued to live in their small second-floor rooms, cooking on a gas stove, because the kitchen had not yet been built.

The home, when finally completed, was a large, three-story, Elizabethan-style house with peaked gables, much larger than the original plans called for. The building was constructed of red brick. The interior, finished in dark wood, had a huge center hall and spacious rooms connected by wide doorways. The rooms included a dining room, butler's pantry, big kitchen, parlor, library, and multiple bedrooms.

The larger bedrooms were on the second floor. The third floor had a series of smaller rooms for servants. The house also featured a large laboratory (the first section of the home to be built and occupied), an office, and several attached greenhouses. The office had built-in shelves for displaying Steinmetz's treasures and collectibles, such as an Indian arrowhead collection, various types of electric light bulbs, and many other odds and ends that caught his attention and engaged his curious, wide-ranging intellect. Because the built-in shelves went from floor to ceiling, a library ladder fitted with wheels went around the room on a track, enabling Steinmetz to reach any item on any shelf.

Why did Steinmetz, a bachelor, build such a large house with many bedrooms? Could it be that he hoped one day soon to fill it with the sound of the laughter and good cheer of family, friends, and children, this man who had vowed never to marry? The size of the home and rooms seemed too big for such a small man, or really any man, to live in by himself. Steinmetz wandered the halls forlornly. Instead of being pleased with his new castle, he found it cold and lonely, and so continued to live in his small bedroom above the laboratory, which was filled with things that were familiar and comfortable: beakers, flasks, tools, copper wire, nuts and bolts, electronic components, machines, and his library of reference books.

Steinmetz said later that he did not like living at home alone in his stately manse, which was a shame, considering the expense and trouble he went to in building it. And as it turned out, he didn't have to. At least not for long.

Here's what happened: The odd couple separated, but only for a brief period. The reason for the separation is that in 1903 Joseph Hayden got married, and Corinne Rost, his new bride, was not keen on the newlyweds living with a strange little man whom she did not know well. At her insistence, Hayden moved out of the Wendell Avenue home, and Joe and Corinne rented a flat on Park Avenue, which was on the other side of town.

Joe and Corinne went on a brief honeymoon. The first night they returned, Steinmetz showed up unexpectedly at their door. Joe naturally asked him to stay for dinner, and Corinne agreed, though somewhat reluctantly. This routine became more frequent until Steinmetz came over for supper almost every day and was a fixture in the Hayden residence.

Finally, Steinmetz suggested to the Haydens that it would make more sense if they moved in with him. Corinne, knowing that Charles and Joe had eaten almost nothing except meat and potatoes virtually every night, said she would agree to the move only if she was in charge of the household. She insisted she would keep the home neat, clean, and free of cigar ash and mess. Additionally, she would prepare nutritionally balanced meals, including vegetables. And whatever Dr. Steinmetz and Joe were doing in the lab, they would stop, come to the table, and eat what she had cooked them for dinner, *when* she called them to dinner. Steinmetz readily agreed to all her terms, and the Haydens moved into the Wendell Avenue home, where they would stay with Steinmetz for the rest of his life.

Mrs. Hayden was not enthusiastic at first. But she knew how fond Joe was of Steinmetz, and her feelings for Steinmetz slowly evolved from tolerance to acceptance and, eventually, to fondness and love.

One factor that made the relationship closer was that, in 1905, Steinmetz legally adopted Joe Hayden, who was at the time age 24. It surely must have struck many as odd: a boss at a big corporation adopts one of his employees as his son. Certainly, in a modern corporation of today, such an intensely personal relationship between manager and employee would raise eyebrows in human resources. But Hayden had come to treasure Steinmetz as a friend, teacher, and mentor, and to both men the adoption seemed natural and normal rather than the oddity it in fact was. From that point on, Hayden always called him "Dad."

And Steinmetz had not just adopted Joseph Hayden. When the Haydens had three children, Steinmetz legally adopted the entire Hayden family, all of whom lived with the diminutive scientist in his grand home with its modern laboratory, a huge greenhouse, and a small zoo. Steinmetz was like a father to Hayden and his wife, and like a grandfather to the Haydens' three children, Joe, Billy, and Marjorie.[11] At last, Steinmetz had what he had always wanted, a family, and he would never be lonely again.[12]

6

Modern Jove Hurls Lightning in a Lab

There are incredibly powerful forces in everyday nature that can cause untold damage to towns, cities, and structures and take many human lives. These include fire, water, wind, earthquakes, volcanic eruptions, and of course electricity, which threatens us as lightning bolts from the sky. Lightning was so feared by our ancestors that in Greek mythology the primary supernatural power of the greatest of the gods, Zeus, is his ability to hurl bolts of lightning.

The earth is struck by approximately 8 million lightning bolts a day.[1] The heat from lightning can reach 50,000 degrees Fahrenheit,[2] and the energy of a lightning bolt can briefly exceed the power of a nuclear reactor. Although the bolt contains hundreds of millions of volts, the flash takes only one-millionth of a second. A lightning bolt generated by a thunderstorm can strike a spot or object as far as 10 miles away.[3] Lightning bolts travel at up to 60,000 miles per second as a jagged discharge of current, about as wide as your finger. Because sound travels much more slowly—about 700 miles an hour—we see the flash first, then hear the thunder a few seconds later, depending in the distance between the lightning strike and where we are.

The U.S. National Weather Service reports that from 1987 to 2016, the United States has averaged 47 reported lightning fatalities a year. Approximately 10 times that many people are actually struck by lightning annually, but 90 percent of those are not killed, though most are injured, often seriously.[4] While severe injury from being hit by a lightning bolt is typical, a few escape it, some miraculously so. In 1980, a 62-year-old man who had become blind as the result of a highway accident regained his sight after being struck by lightning near his home.[5]

Naturally, as a scientist whose primary area of investigation was electricity, Steinmetz was fascinated by lightning. And by sheer happenstance, he had an opportunity to study lightning and unravel its mysteries afforded to no other scientist.

Steinmetz had a summer house on Viele Creek near Lake Mohawk, where he spent much time with friends and family. In particular, he liked to float in a canoe on the lake, or more specifically in the creek. He placed a board across the width of the canoe to serve as a desk and happily sat drifting on the tranquil water in the great outdoors, working on his mathematical and engineering problems. When he told his family he was going fishing, they knew that meant sitting in the canoe thinking, but not fishing, because he never took a fishing rod with him.

The summer house was a rustic cabin perched on a bluff 50 feet up overlooking the water. It had a large main room, several bedrooms, and a gabled roof. Steinmetz christened the property "Camp Mohawk."

One day in the summer of 1920, there was a thunderstorm. A bolt of lightning struck the lake house. The lightning shattered a bedroom window, flew across the room, and struck and blew apart a large mirror hanging on the wall, blasting it into many pieces.

Ever curious, Steinmetz collected all the pieces of mirror glass. Then, he spent a weekend putting them back together, much like solving a jigsaw puzzle. When examining the mirror fragments, Steinmetz noticed that the mirror's amalgam, an alloy of mercury on the back of the glass, had been partly burned, which produced patterns of markings. Matching up the markings aided Steinmetz in fitting the pieces together correctly.

But reassembling the mirror was more than just a way to pass the time, as solving a jigsaw puzzle usually is: Once the mirror was glued together and made whole, Steinmetz carefully studied the pattern of the cracks in the glass. Why? *Because they revealed to him the electrical path that the lightning discharge had taken!*

Once the pieces were fitted together correctly, Steinmetz next had the mirror placed between two pieces of plate glass. The whole assembly was sealed along its edges, so that the plate glass held the reassembled mirror firmly in place, with no pieces shifting position or falling out. Once sealed, the mirror was transported from the camp to the laboratory in Schenectady. There, within his well-equipped electrical workshop, he began in earnest to study the patterns of the lightning strike, to learn all he could about the electricity that comes to us from the clouds—and, specifically, the path the electricity in a lightning bolt takes.

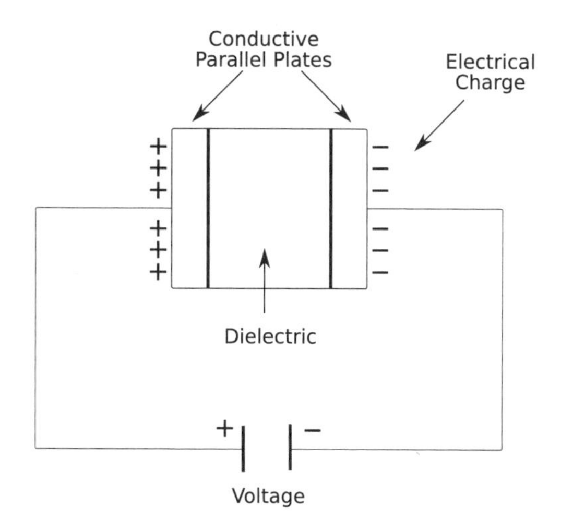

Fig. 6-1. A condenser.

The Steinmetz lightning machine

Because he could not very well count on another outlier like lightning coming through his bedroom window and leaving its signature on the wall, Steinmetz decided he needed to build a machine that could create artificial lightning on demand. Steinmetz constructed his first indoor lightning generator by building a frame of vertical wooden posts with crossbeams between them, housed in a lab about the size of a football field. To this frame he attached the following:

- 200 coated glass plates arranged in groups of 50 in two banks
- glass bulbs of varying shapes, which functioned as rectifiers
- a network of wires

How did the lightning machine work? The glass plates served as "condensers." Condensers store electric current, which they take on gradually, much as clouds become charged with electrons. Also called capacitors, condensers consist of two charged metal plates or other conducting surfaces, one positive and the other negative. The plates are separated by a dielectric, which is essentially an insulator (fig. 6-1).

Steinmetz made a mental leap: he realized that condensers filled with electric current beyond their capacity could behave like clouds with an excessive supply of electrons. Filled to overflowing with negatively charged particles, the condensers in Steinmetz's lightning machine discharged the excess electricity as artificial lightning bolts. These discharges were much like natural lightning, only on a smaller scale.

Steinmetz's first lightning machine discharged 10,000 amperes at around 100,000 volts. The energy of the miniature lightning discharge, which lasted only a hundred-thousandth of a second, was a million horsepower. The energy of a real flash of lightning is 500 million horsepower and 50 million to 100 million volts. Lightning discharges its enormous voltage in the blink of an eye. The resultant rapid release of electrical energy gives lightning its great power and explosive force. Objects hit by lightning are usually scorched, burned, and sometimes blown apart.

Often the world is slow to recognize major scientific and technological breakthroughs. To hasten the process of acceptance, the inventors themselves, their universities, companies, or the government decide to test the new technology on a large scale and in a dramatic way. So in 1922, Steinmetz constructed a miniature model village out of wood, similar in size, scope, and appearance to the tiny towns that model railroad enthusiasts set up around the train tracks.[6] Then he built a 120,000-volt lightning generator and positioned it above the model town.

He announced that there would be a demonstration in his lab of a machine that makes artificial lightning. He invited members of the general public and the press. When they were seated, he darkened the room and flipped the generator switch. Charge built in the condensers, and shortly thereafter the generator hanging from the ceiling produced a miniature lightning storm that destroyed the model village in rather spectacular fashion.

But the demonstration was not just for show. The artificial lightning machine gave Steinmetz and other scientists a better understanding of how lightning behaved. This, in turn, helped engineers develop more effective devices for protecting buildings, other structures, and electronic equipment from lightning bolts.

Lightning caused problems in the power lines of the electrical grid that was expanding across the country in the early 1900s. Direct lightning strikes or nearby flashes created destructive power surges on transmission lines. To counter this, engineers designed several protective devices, including heavy line insulators, lightning arresters, and power-limiting reactors.[7]

Now, with Steinmetz's lightning machine, experimental lightning arresters could be tested at will in the laboratory instead of waiting for lightning to strike. The increased testing ability under laboratory-controlled conditions helped make the development of lightning protection technology both safer and less expensive. And in the controlled environment of the laboratory, tests of lightning arresters were also more precise as well as easier to conduct than any previous tests.

In many of the tests, the artificial lightning generator produced bolts that struck lightning arresters that were in turn connected to the circuits of an actual building, resulting in a more accurate determination of the degree of protection the arrester afforded. Because of this improved trial-and-error testing, performed with artificial lightning that could be generated on demand at any time, without waiting for a thunderstorm, engineers were able to develop more reliable protection for electrical transmission systems, generating stations, and building circuitry that was far more effective and sophisticated than Ben Franklin's early lightning rods.

Designs of lightning arresters

Ben Franklin invented the first lightning rod in 1749. (In Ray Bradbury's 1962 novel *Something Wicked This Way Comes*, Tom Fury is a lightning-rod salesman.) Franklin got the idea for the lightning rod from electrical experiments he performed in which he found that electricity was attracted to sharp needles; from this he came to the conclusion that electricity is attracted by points. And if lightning had this property, then it could be attracted by points in lightning arresters or, in Franklin's design, lightning rods that could attract lightning away from buildings, thereby protecting them from damage. Franklin told a colleague (italics mine):

> There is something . . . in the experiments of points, sending off or drawing on the electrical fire, which has not been fully explained . . . for the doctrine of points is very curious, and the effects of them truly wonderful and from what I have observed on experiments, *I am of the opinion that houses, ships, and even towers and churches may be effectually secured from the strokes of lightning* by their means; for if, instead of the round balls of wood or metal which are commonly placed on the tops of weathercocks, vanes, or spindles of churches, spires, or masts, *there should be a rod of iron eight or ten feet in length, sharpened gradually to a point like a needle*, and gilt to prevent rusting, or divided into a number of points, which would be better, the electrical fire would, I think, be drawn out of a cloud silently, before it could come near enough to strike; and a light would be seen at the point.[8]

The effect that takes place when electricity from the atmosphere builds up around an insulated piece of metal is often referred to as a glow discharge or brush discharge. The electrification causes the oppositely charged electricity within a metal rod to come to the top of the rod. If the difference in charge between the top and the body of the metal rod is significant, electricity discharges upward to neutralize the opposing force, creating electrically charged gas molecules known as ions.[9]

Here's how Franklin described his invention, the lightning rod, in an article in *Poor Richard's Almanac* in 1753 (italics mine):

> It has pleased God in His goodness to mankind, at length to discover to them the means of securing their habitations and other buildings from mischief by thunder and lightning. The method is this: Provide a small iron rod (it may be the rod-iron used by the nailers) but of such length that *one end [is] three or four feet in the moist ground, the other may be six or eight feet above the highest part of the building.* To the upper end of the rod fasten about a foot of brass wire the size of a common knitting needle, sharpened to a fine point, the rod may be secured to the house by a few small staples. If the house or barn be long, there may be a rod and point at each end, and a middling wire along the ridge from one to the other. A house thus furnished will not be damaged by lightning, it being attracted by the points and passing through the metal without hurting anybody. Vessels, also, having a sharp-pointed rod fixed on the top of their masts, with a wire from the foot of the rod reaching down, round one of the shrouds, to the water, will not be hurt by lightning.[10]

An improved lightning arrester was developed by Steinmetz's lab assistant and adopted son, Joe Hayden, collaborating with E. E. F. Creighton, F. W. Peek, and others. Their arrester worked by capturing and storing the energy of capacity and inductance.

The aluminum arrester was built from a series of cells. The cells were shaped like cones and stacked into aluminum electrodes with an electrolyte consisting of salt in solution. Aluminum resists corrosion, so neither the salt nor its ions interfered with the performance of the aluminum arrester. An alternating current passed through the aluminum cells, in which the electrodes were coated by a thin nonconducting film of alumina, a type of aluminum oxide. The film on the electrodes became thicker with repeated usage, until it could hold back the lightning's extreme voltage.

Sometimes extreme voltage punctured the aluminum oxide coating. But these punctures were soon refilled as the current passing through them built up the alumina coating again and closed the holes. In this way, the aluminum cell served as a self-repairing electrostatic condenser, with a

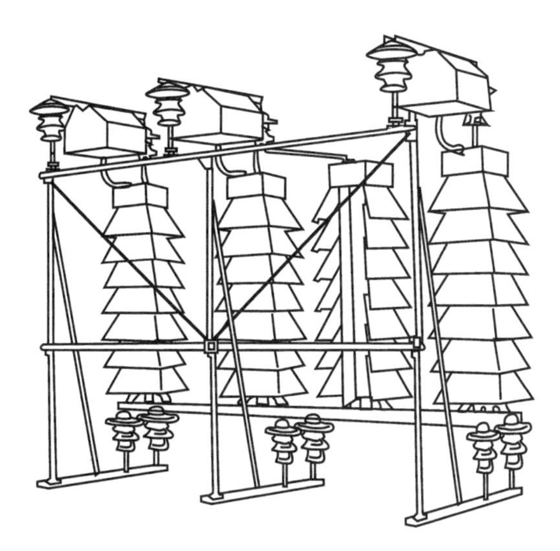

Fig. 6-2. Oxide film lightning arrester.

disruptive strength equal to the voltage the device was designed to protect machines and building circuits from—about 250 to 300 volts per cell.

The aluminum arrester acted against the high-voltage discharge of lightning with a counter electromagnetic force equivalent to the normal circuit voltage. The lightning discharge passed through the arrester, which short-circuited the current of the lightning voltage spike. The machine being protected by the arrester therefore was not disrupted or damaged.

Hayden's aluminum cell arrester could discharge continual disturbances, with overvoltage oscillations occurring at every half wave, for up to several hours. Beyond that, the arrester would overheat from the temperature rise caused by the accumulated energy of these high-voltage lightning strikes.

Aluminum cells became the most widely used lightning protectors. However, they required maintenance on a frequent basis: the aluminum arresters had to be charged daily and kept filled with liquid electrolyte. They were difficult to test without taking them apart, except by watching the appearance of the charging arc, or measuring the charging current.

For this reason, electrical engineers developed a new and improved type of lightning arrester: the oxide film lightning arrester. It had all of the characteristics and advantages of the aluminum cell arrester. But the oxide film model did not require any charging or special attention, contained no liquid electrolyte, and like the aluminum cell arrester could be located outdoors as well as indoors (fig. 6-2).

The oxide film arrester acts like a counter electromagnetic force equivalent to the normal circuit voltage. It freely discharged any excessive voltage but held back the normal machine voltage. Therefore the discharge was limited to the energy of the overvoltage, as in the aluminum arrester. Like the aluminum arrester, the oxide film arrester could continually discharge recurrent surges, such as arcing grounds, without endangering itself for a considerable time, sufficiently long enough to eliminate the disturbance.[11]

Steinmetz had another reason for devoting himself to generating and experimenting with artificial lightning. He knew lightning bolts had tremendous power. So Steinmetz reasoned that if he could find some method of preventing the lightning strikes from disrupting power transmission cables, current insulating technology would be more than sufficient for keeping generator currents where they belong.

And here's the problem he found with high voltages produced by power generation stations: When the transformer steps up the voltage for long-distance power transmission, the force in the wire becomes too great. As a result, the electricity escapes from the wire. Knowing that, Steinmetz experimented with different voltage levels and insulating materials. Insulation is a material, such as rubber, that has resistance so great that it restricts the flow of electric current to the conductor and prevents the electricity from escaping to the surrounding space or any object touching the insulated wire.

Eventually Steinmetz found the optimal voltage and insulation to completely contain the electrical energy within the copper wire.[12] Plus, with this heavy insulation shielding the transmission lines, any lightning that hit the power system would be deflected by the insulation and jump harmlessly to the ground instead of damaging the equipment.

As time went on, Steinmetz became interested in making his artificial lightning machines bigger and more powerful. After the dramatic destruction of his model village by artificial lightning inside his laboratory, he built a lightning tower outdoors on the General Electric campus using 120,000-volt generators. He was dubbed "forger of thunderbolts" by the press. Some science historians consider artificial lightning to be his greatest invention.[13]

After Steinmetz's death, the GE exhibit at the 1939 New York World's Fair included a display, dubbed Steinmetz Hall, equipped with a pair of Steinmetz's artificial lightning machines. The twin generators discharged 10 million volts of electricity across a 30-foot gap at regular intervals, each bolt accompanied by a loud thunderclap. The Steinmetz Hall exhibit drew as many as 22,000 visitors daily.[14]

7

Driven by Electricity

In the twenty-first century, Elon Musk is credited as being a pioneer in electric cars. His electric car (fig. 7-1) and the company that makes it, Tesla, are named after Nikola Tesla, Steinmetz's chief rival for the honorary title of greatest electrical wizard of the twentieth century. And today Musk and Tesla get the lion's share of the enormous publicity about electric cars, which the press reports on as if it were a new idea. But it is not: electric cars have been around since the nineteenth century. And Charles Steinmetz was at least as involved in this technology as Nikola Tesla.

Electric cars are in many ways similar in design and function to conventional cars:[1] they have a body, chassis, wheels, brakes, accelerator, and transmission (however, the transmission in an electric car has only one gear). From the outside, electric cars and normal cars look pretty much alike, except electric cars have no exhaust system or pipe because they do not produce emissions. Other than the absence of a tailpipe, the main difference is that the gasoline engine is replaced by a much quieter electric motor. The three most common types of motors used in electric cars today are DC brushless, AC induction, and permanent magnet motor.

The electric motor is attached to a controller (described below), which gets its power from an array of rechargeable batteries. The batteries are located either under the car or in the trunk. In a conventional automobile, one 12-volt lead storage battery that provides power for the ignition to start the engine as well as for interior and exterior lights, radio, and climate control. Electric cars have multiple batteries that actually provide the power that moves the vehicle. A regulator on the batteries ensures that the amount of energy produced by the batteries and consumed by the motor is

Fig. 7-1. Modern Tesla electric car.

constant, which prevents batteries from burning out. Instead of a gas gauge alerting you when fuel is low, the electric car has a voltmeter telling you when it is time for a recharge.

The batteries produce direct current. The controller converts the direct current into the polyphase alternating current required by the electric motor. The electric motor converts the electrical energy to mechanical energy, which propels the car forward. Variable potentiometers are controlled by the accelerator pedal and determine how much current is delivered from the controller to the motor.[2]

The earliest electric cars, most of which were small, were designed and built in the first half of the nineteenth century by a number of independent inventors in many different countries, including the United States, Hungary, the Netherlands, France, and England. These cars were mainly experimental and not intended for mass production or commercial sale.

These smaller models were curiosities, underpowered and slow, and could transport only the driver and maybe one passenger. The first larger electric cars, capable of carrying both a driver and multiple passengers, were built in the second half of the 1800s. In the United States, chemist and inventor William Morrison constructed an electric vehicle that could hold six passengers and reach a top speed of 14 miles an hour. By comparison, the world's fastest human being as the time of this writing, Usain Bolt, can briefly attain a running speed of nearly double that—almost 28 miles an hour.[3]

Other inventors and small companies in the United States began building their own electric cars, including a fleet of more than five dozen electric taxis in New York City. And electric cars were not an oddity or an outlier in the early days of the automotive industry; by 1900, a third of all cars on the road were powered by electricity.[4]

With the success of the electric car, transportation engineers turned their attention toward developing railroad cars driven by electricity. Steinmetz publicly stated his belief that electricity would replace steam in locomotives. His reasoning was that steam cost more and was less efficient. Steam engines slowed on upgrades, while an electric motor would maintain full power and velocity on an upward incline.

Although Steinmetz sometimes consulted on electrified railroad projects, it was never a big part of his work. However, starting in 1912, he became more actively involved with designing and building electric vehicles for the road. But he was more interested in electric trucks than electric cars. Why? He said that because trucks were used for business purposes, customers would be more willing to pay a premium price for an electric truck than they would for an electric car.

In the early 1900s, the overwhelming majority of trucks had internal combustion engines fueled by gasoline, although some electric trucks were in service. Steinmetz's calculations showed him that while companies would continue to use gasoline-fueled trucks for long-haul transportation, for shorter trips the electric truck would prove more economical. Because gasoline was expensive for that era (7 cents a gallon[5]), Steinmetz stated that the use of electric trucks would reduce the cost of delivering freight. Businesses would save money on transportation, which he believed they would in part pass on to customers in the form of lower prices for their goods.

Steinmetz was not the first to design and build an electric truck, but he improved on the technology. In some areas of his work, such as the magnetite arc lamp and the mathematics of hysteresis, Steinmetz advanced the state of the art in technology or theory considerably. But there was nothing revolutionary or particularly innovative in his design for an electric truck. The main difference was that the Steinmetz truck carried a larger number of storage batteries than competing brands. A good idea? Yes. But his electric truck was powered by existing, off-the-shelf battery technology; he did not design an improved electric battery or a better way to power the vehicle.

Steinmetz had practical experience with electric vehicles already: he owned and drove an electric car made by Detroit Electric Car. Ironically, the car was powered by storage batteries made by his rival, Edison. Years later, Thomas Edison and Henry Ford teamed up to build an electric car, but their joint effort was not successful.[6]

In 1920, Steinmetz formed the Steinmetz Electric Motor Car Company to design prototypes of several electric vehicles, including both trucks and cars. The company was in Brooklyn, where by 1922 it had produced an industrial truck, followed by a lightweight delivery car and a five-passenger car for consumers. Steinmetz had stated that an electric vehicle would outpower a steam vehicle on hills. For publicity, he invited spectators to watch his truck's first-test drive on a steep hill in Brooklyn. As Steinmetz had predicted, the electric truck was indeed powerful and did well on the hill.

Steinmetz was primarily an engineer, not an entrepreneur. Unlike Edison and Westinghouse, but much like Tesla, he seemed to have little aptitude for business. So in the early 1920s, he accepted the offer of a group of businessmen to help him in his motor vehicle venture. These businessmen became either partners or investors (it is not clear which) in the Steinmetz Electric Motor Car Company. Their intention was to produce 1,000 electric trucks and 300 electric cars a year.

Because the Steinmetz electric vehicles used standard batteries and parts, and not a special Steinmetz invention, the businessmen didn't really need Steinmetz to run the company, though he added considerable value, both as a technical advisor and for the publicity lent by his fame as an electrical wizard. He was given an honorary position and an office at the company plant. The executives began offering shares in the new Steinmetz venture, and investors eagerly forked over cash to buy stock in the new company, so they could get in on the ground floor of what everyone thought would be another Steinmetz breakthrough.

Unfortunately, his partners were not the business wizards they had represented themselves to be. The company floundered for a couple of years, and when Steinmetz passed away in 1923, the Steinmetz Electric Motor Car Company closed its doors. As a result, Steinmetz electric trucks and cars were never mass produced. In addition, the stock price floundered, and investors felt they had been cheated.[7] By comparison, investors who

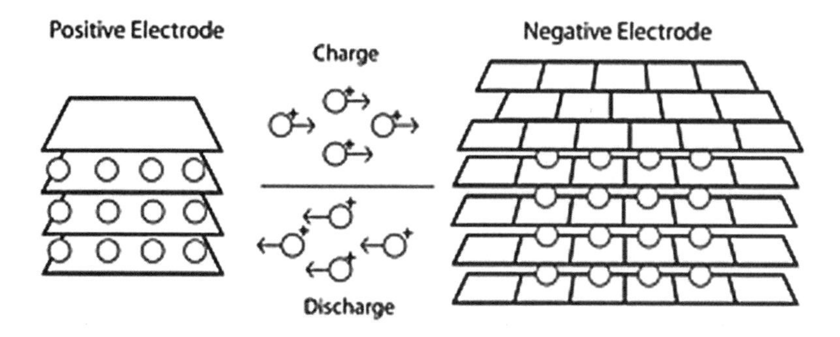

Fig. 7-2. Lithium-ion battery.

bought stock in Tesla Motors saw their shares rise in price from November 2012 to February 2017 by a hefty 977 percent.[8]

Today, people wrongly perceive electric cars as an innovation, using some mysterious new power source. In point of fact, today's Tesla gets its electric power from batteries, just as the Steinmetz truck did. The major difference is that today the Tesla runs not on regular lead-storage car batteries but on lithium-ion batteries.

The Tesla battery uses essentially the same lithium-ion technology as the batteries in your smartphone, tablet, smartwatch, or laptop. The only difference is that the Tesla's lithium-ion battery is far bigger and more powerful (fig. 7-2), which of course is needed to power a car weighing over two tons.[9]

In a lithium-ion battery, the electrolyte transports positively charged lithium ions from the anode to the cathode. While the battery is discharging and providing current to a car or other device, the release of lithium ions to the cathode generates a flow of electrons.[10] Each Tesla has a central battery made up of thousands of lithium-ion cells with a total combined weight of about half a ton. The Tesla comes equipped with a heating system that enables the car battery to function in cold weather.[11] The Tesla Model S can go 350 miles or more on a single charge.[12] The Tesla Powerwall battery, which consists of an array of smaller connected cells, is used in their consumer models and weighs 276 pounds.[13] In 2018, Tesla CEO Elon Musk spent $6 billion to build a massive battery factory, which is the size of 95 football fields, in the Nevada desert. It can make half a billion batteries for electric cars a year.[14]

8

An Exaltation of Geniuses

Almost everything in the world—from scientific discoveries and art to puddings and apple pies to movies and video games—exists on a bell curve (fig. 8-1). The middle of the bell curve represents mediocrity, which characterizes the majority of things or people. Most of everything, from restaurants and kitchen knives to plays and poetry, is just average, or a little better than average, or a little worse.

On the far left side of the curve is the small minority of things in any category that are of especially inferior quality: lousy teachers, shoddy home construction, cheap wine, and bad pizza. On the far right side of the curve is another minority: the things in the category that are in the top 10 percentile, the best of the best. In champagne, Dom Pérignon. In steak, filet mignon. In the category of science, Steinmetz was at the extreme right of the bell curve.

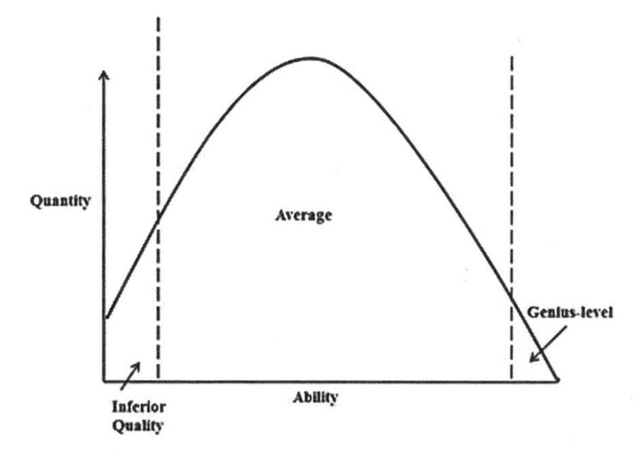

Fig. 8-1. Bell curve.

The bell curve is consistent with Sturgeon's law, a theorem devised by science fiction writer Theodore Sturgeon, which said that only 10 percent of everything—pies, cars, plays, music CDs, detergent brands, TV commercials, and anything else you can think of—is really good; the other 90 percent ranges from mediocre to lousy.

Sturgeon's Law holds for people too: their behavior, personalities, intelligence, and skills. In engineering, only 10 percent of those in the profession are really, really good. The other 90 percent are mediocre at best or incompetent at worst. The same holds for musicians, auto mechanics, architects, plumbers, and every other profession, hobby, avocation, skill, and activity. As an old joke notes, 50 percent of all doctors graduated in the bottom half of their class in medical school, yet many of those doctors have plenty of patients.

Now, as it happens, in almost every field of human endeavor—from boxing and baseball to opera and rock to science and engineering—the superstars in a given activity, industry, or pursuit more often than not know and associate with many of their fellow luminaries, whether through casual meetings, correspondence, or friendship (or in the twenty-first century, via Skype, email, and social media). Steinmetz was no exception.

You might guess that being a 4-foot-tall hunchback with a curved spine, Steinmetz would be shy and retiring. But on the contrary, he was friendly, outgoing, and a good conversationalist, and enjoyed meeting and spending time with others who shared his scientific and technical interests.

Scientists are often wrongly portrayed as reclusive introverts, toiling alone in a dimly lit laboratory or in front of a chalkboard to make their great discoveries. In the early days of modern science, up until the nineteenth century or so, it seemed to the general public that a lot of the great contributions to science, technology, and mathematics were made by individuals—lone wolves—and not teams. The brilliant men and women of science, mathematics, and engineering from the eighteenth, nineteenth, and early twentieth centuries—from Marie Curie and Sir Isaac Newton to James Maxwell and Joseph Priestly—are almost always pictured performing experiments alone at the workbench.

In early horror and science fiction movies, the "mad" scientist was stereotyped as a genius always cloistered in his lab alone, or with a single assistant, an "Igor," at a workbench surrounded by flasks and beakers bubbling and giving off vapor (a special effect achieved by putting a chunk of dry ice in warm water).

In reality, however, scientists almost never achieved their greatest triumphs without collaboration. Steinmetz had his chief lab assistant Joseph Hayden and regularly shared ideas with Edison, Einstein, engineers at GE and the college, and other contemporaries. As Yale professor Priyamrada Natarajan writes: "Although advances in science and technology are often portrayed as the work of solitary men—for example, Isaac Newton, Thomas Edison, and Albert Einstein—science has always been a collective enterprise, dependent on many individuals who work behind the scenes."[1]

To be sure, in the modern industrial and academic world of the twenty-first century, the vast majority of breakthroughs in science and technology, and especially those in industry, are now achieved in teams. Almost all engineers, scientists, and programmers work in teams, and it is these teams that solve problems. While there is a team leader, the members are equals, or at least somewhat peers. There is no lone genius with lackeys doing the grunt work.

A survey by ClearCompany confirms that today three out of four workers rate teamwork as very important.[2] Another survey, The Ernst & Young 2016 Global Private Equity Survey, found that more than six out of ten investors rank team stability as their biggest concern about companies they might invest in—more important than operational excellence or strategy.[3]

In this chapter, we take a look at some of Steinmetz's associates with whom he shared companionship, conversation, time, and ideas.

Guglielmo Marconi (1874–1937)

In 1922, Guglielmo Marconi visited Steinmetz at General Electric headquarters in Schenectady, New York, and also at his Wendell Avenue home there.[4]

Marconi is best known for developing shortwave radios used for signaling. These radios used the electromagnetic waves first discovered by Heinrich R. Hertz (1857–1894). Marconi's early shortwave radios could only communicate via Morse code, a signaling method developed by Samuel Morse for telegraph communication.

The telegraph is a simple device. An operator taps a button, which causes a metal stylus to make contact with a metal plate. When the contact is made, a signal is sent. A quick tap sends a short signal burst, which came to be known as a dot, while holding the key down a bit longer produces a signal of slightly longer duration, which was called a dash. Telegraph

Guglielmo Marconi.

messages were sent in Morse code, a simple code of Morse's own invention. Each letter in Morse code (fig. 8-3) is represented by some combination of dots and dashes. The phrase in Morse code that has become iconic in our culture is three short for S, three long for O, and another three short for S, which is SOS, and stands for "save our ship"—a distress signal sent when a vessel at sea was sinking

The first working telegraph system, built in 1844, enabled Morse code communication between two cities separated by a distance of about 40 miles, Baltimore and Washington, DC. During the Civil War, 15,000 miles of telegraph cable were laid for military purposes.[5]

In 1899, Marconi boosted the range of his shortwave radio so that Morse code signals could be sent across the English Channel. The America's Cup yacht race began to use Marconi's shortwave radio for ship-to-ship communication. In 1901, Marconi transmitted a shortwave radio signal across the Atlantic Ocean from England to Canada. And in 1932, Marconi's company won a contract to establish shortwave radio communication between England and other countries in the British Empire.[6]

You might wonder how this is possible. Wouldn't the curvature of the earth over such a distance make the signal miss the receiver and go off into outer space? Well, the signal *did* beam up toward space. But the radio waves were reflected off ionized layers in the upper reaches of the atmosphere.

In Marconi's radios, the electromagnetic waves were detected on the receiving end by a container filled with metal filings. When the electromagnetic waves from Marconi's radio struck the receiver, it generated an electric current, as detectable as the closed and open circuits on Morse's telegraph.

(Two fun side notes: By the time Thomas Edison came to Schenectady to visit Steinmetz at the GE plant, Edison was quite deaf. When Steinmetz

Fig. 8-3. Morse code.

saw that Edison could not hear him, he found an immediate solution: he tapped out what he wanted to say on Edison's knee in Morse code![7] When Marconi came to Schenectady to visit Steinmetz at his Wendell Avenue home, he looked around the by-then-famous menagerie (see chapter 9) and asked, "Where is your Gila monster?" Steinmetz sadly reported that his beloved pet had died after becoming lethargic and becoming unable to eat.)

During World War I, Marconi invented a radio beam pilots could use to fly by instruments alone, even if they had no visibility from the cockpit. A couple of years later, physicist Reginald Fessenden figured out how to modulate radio waves so they could send not just Morse code but also transmit voices and eventually music. In 1909, Marconi won a Nobel Prize in physics.

Henry Ford (1863–1947)

Many famous scientists and industrialists considered Steinmetz a friend and colleague and came to him for his technical knowledge. Henry Ford once paid him $10,000 for just two days of consulting on fixing a large generator in the Ford plant, an outrageously exorbitant fee for that era.[8]

Here's the story: The Ford plant was having a generator problem, and the plant's electrical engineers were unable to fix it. So Ford called in Steinmetz, whom he knew to be the best in his field. When Steinmetz arrived, he listened to the faulty generator as it ran. Then, he went off and made extensive calculations. He was given use of a small office and a cot so he could rest when weary.

Henry Ford.

Upon emerging from the office after two days of working on the problem, Steinmetz walked over to the giant generator and, with a piece of chalk, drew an *X* on its side. "Here's your problem," he told Ford's plant engineers. He instructed them to remove the metal plate upon which he had make his chalk mark. Next, he advised them to replace 16 windings from the field coil. As soon as they did, the generator began working perfectly—problem solved.

As the story goes, Henry Ford was thrilled until he got an invoice from Steinmetz in the amount of $10,000. Ford acknowledged Steinmetz's success but balked at the charge. He asked for an itemized bill, which he received from Steinmetz as follows:

> Making chalk mark on generator: $1.
> Knowing where to make mark: $9,999.
> Total: $10,000.

Ford Model T.

Of course, Ford paid the bill. Not doing so would be terrible publicity and would mean Steinmetz would never come to Ford's aid again if the need arose. And it did. Another time, Ford had a problem with his flagship car, the Model T: the headlights were bright enough when the car was being driven but dimmed when the motor was idle.

Ford visited Steinmetz at his home, and the two men went into Steinmetz's home office to discuss the problem. Just then Midge, Steinmetz's adopted granddaughter, opened the door and asked Steinmetz to read her a bedtime story, as was their custom.

Steinmetz got up and told Ford that he had to read Midge her bedtime story and that he would be back in half an hour. Ford, already a successful industrialist and an enormously wealthy and powerful man, was furious. How dare Steinmetz treat him this way! Every minute he sat in the office waiting for Steinmetz to return, he grew angrier and more irritated. But when Steinmetz returned, he took out a pad of paper and quickly sketched a solution he had devised for the headlight problem. And suddenly Ford was irritated no longer.

Ford and Steinmetz are emblematic of the two types of inventors, engineers, and scientists: the obsessively focused achiever versus the intellectually curious Renaissance man. Ford was the former, driven to change the face of U.S. transportation, manufacture cars that were affordable to the average man, build a great company, and in the process make himself rich beyond the dreams of avarice. All of Ford's energies were focused on these goals, and he had time for little else.

Once when Ford was testifying in court, an attorney attempted to prove that Ford was not a loyal and patriotic American. He asked Ford many simple questions about U.S. history and government, the answers to which were common knowledge but which Ford could not give. When he then asked Ford how an American could not know these things, Ford became irate.

"I don't know them because I don't need to know them," he replied. "Why should I fill my brain with useless knowledge, when I have a staff of assistants at my beck and call who can get these facts for me?"

There are many people today who agree with Ford. Comedian Steve Martin once attributed his success to the fact that for eight years he dedicated himself almost wholly to perfecting his stand-up comedy, to the exclusion of all else. In his 2014 book *Essentialism* (Crown Business), Greg McKeown also agrees with this focused approach, saying, "There are far more activities and opportunities in the world than we have time and resources to invest in. Only once you give yourself permission to stop trying to do all can you make your highest contribution toward the things that really matter."

On the side opposite from the essentialists are the Renaissance men and women. They are interested in many things if not most things or even everything. Their intellectual curiosity is boundless. And despite what McKeown says, many of our greatest achievers were Renaissance men and women.

Leonardo da Vinci is perhaps the most famous Renaissance man: a painter, sculptor, illustrator, and scientist who was successful in many pursuits, including anatomy, architecture, engineering, geology, hydraulics, inventing, and the military arts.[9] Steinmetz, too, represented both worlds: he was a hardworking superachiever and a Renaissance dabbler. He was intensely focused on his core activity of electrical research, which yielded some of the greatest technological innovations contributing to today's modern world. His primary study was in math and electricity, but he read widely at the same time, and kept an active interest in his many hobbies and interests, which ranged from photography and zoology to botany and politics.

Interestingly, Ford played a small role in Aldous Huxley's 1932 novel *Brave New World*, a book that envisions a future society in which the calendar dates are based not on the birth of Jesus Christ but on the invention of the Ford Model T. Dates later than the invention of the new car were designated AF for After Ford.

Albert Einstein (1879–1955)

Einstein is best known for his theory of relativity, which revealed the amazing fact that time is not constant but can run at different rates depending on your velocity: the faster you are moving, the more slowly time passes for you. And Einstein's most famous equation, $E = mc^2$, which revealed that mass could be converted to an enormous amount of energy, led to both nuclear power and the atomic bomb. He won the Nobel Prize, but not for formulating his famous equation or for developing the theory of relativity. His prize was for his discovery of the photoelectric effect, which showed that light striking metal can cause that metal to eject electrons from its surface.

Albert Einstein.

Einstein met Steinmetz during a visit to General Electric in 1921. Although their work seemed unrelated at the time, today GE is using Einstein's theory of relativity to synchronize power distribution and prevent outages.[10]

You may ask what relativity could possibly have do with electrical power distribution.[11] Well, in the twenty-first century the electrical grid has become increasingly digitalized and connected to the internet. For at least a decade, many utilities have been using two-way communications technology to support remote grid operations.[12] This technology is known as the "smart grid." In the modern digital power grid, the time the current leaves the power station and also the time when it arrives at its destination are both measured with precision clocks, which are synchronized to one hundred millionth of a second. When clocks on the sending and receiving end of the current transmission are perfectly in sync, the result is an accurate measurement of the speed of AC current transmission.

Among other advances, the precision of the digital grid permits utility operators to more accurately predict power outages. Digital power grids can also enable internet-connected services such as smart water networks, electric vehicle chargers, bike kiosks, smart parking, weather and pollution sensors, and smart streetlights and traffic controls. Echelon, an internet-of-things (IoT) company, is a leading provider of smart-connected streetlights.[13] Estimates are that by 2020, 50 billion devices will be connected by the IOT.[14]

To achieve such precision time measurement, the digital clocks are connected to the Global Positioning System (GPS), which in turn is connected to satellites, which orbit the earth at high speeds. In keeping with the theory of relativity, time on these fast-moving satellites slows down ever so slightly compared to time on the earth's surface, making the GPS-connected clock run just a tiny bit more slowly. And so the relativistic difference in the satellite clock's time versus the time of a stationary clock on Earth becomes significant: because they measure time to the hundred millionths of a second, even a slight relativistic effect can throw off the measurement. In the digital energy network, calculations using Einstein's equations from the theory of relativity must be made via high-speed computer. These calculations reconcile the relativistic time difference and enable accurate measurement for the time it takes current to travel from its source to its destination.

In skill and practice, Steinmetz was more of a hybrid mathematician, physicist, and electrical engineer than a pure electrical engineer, and as part-physicist he was so engrossed with Einstein's theory of relativity that he gave several lectures on it. Here is what Steinmetz said in his lectures about the importance of relativity:

> The theory of relativity developed by Einstein and his collaborators is the greatest scientific achievement of our age.
>
> The layman is therefore fully justified in wishing and asking to know what it is about and in his desire to get at least a glimpse of the new and broader conception of the universe and its laws which this theory is giving us and to understand what fundamental revolution in our scientific world conception it is causing in bringing order and system out of the previous chaotic state.

Unfortunately, the relativity theory is intrinsically mathematical, and it is impossible to give a rigidly correct and complete exposition of it without the extensive use of mathematics. The best that can be done, therefore, in explaining the theory of relativity to the layman, and to the engineer who is not an expert mathematician, is to give by analogy, example, and comparison a general conception of the theory and its fascinating deductions and conclusions.

Such a conception must inevitably be approximate only and cannot be rigidly correct. This must become evident to the mathematical physicist. However, it is the best that can be done, and I believe it is sufficient to justify fully the little effort required from the layman to follow the exposition. After all, the non-mathematician is not interested in rigidly following the intricacies of the mathematical reasoning involved.

Rather it is his desire to get a general knowledge and understanding of the new ideas on time and space, on the laws of nature and the characteristics of our universe, which the relativity theory has deduced, and of the wonderful researches into the nature of space which nearly a century ago were carried out by the great mathematicians and have now at length become of physical significance and indeed been the mathematical foundation on which the theory is built.[15]

Thomas Edison (1847–1931)

When Thomas Edison was a boy in elementary school, a teacher sent him home with a note. He was told not to read it and to give it directly to his mother.[16] When his mother read the note, she told her son: "The teacher says you are so smart that the school is not able to provide you with an education adequate for a person of your gifts. Therefore, they advise that I teach you at home." And so Edison's mother kept him home, where he was home-schooled by her; the boy also acquired knowledge as an autodidact.

After his mother died, Edison found and opened the note. The teacher in fact had written: "That Edison boy is addled. He can't learn." Edison rapidly proved his teacher to be wrong at best and a fool at worst. He began working hard both on his self-education and what would be the first of his many money-making ventures. At age 12, Edison sold newspapers; at 15, he worked as a telegraph operator.

Like Edison's mother, Steinmetz's father was also told that his young son was not capable of learning. When the boy was 8 years old, the headmaster of the school proclaimed that young Steinmetz was retarded. But that changed seemingly overnight, and by age 10 Karl Steinmetz was the top student at the school.[17]

Thomas Edison.

Edison received his first patent in 1869 for an electronic vote recorder. In 1870 he opened a machine shop and laboratory in Newark, New Jersey. Five years later he closed this and built the world's first industrial research laboratory in Menlo Park, New Jersey.

Interested in Alexander Graham Bell's new telephone, Edison patented several improvements to the phone in 1877 and 1878. While working on the telephone he noticed that his voice made a needle point vibrate. This led to the phonograph patent in 1877. Edison used the phonograph (fig. 8-8) to record both music and the spoken word. He once told a newspaper reporter that in a few years there would be no printed college textbooks, as they would all be recorded on cylinders.

He was both wrong and right. Paperbound textbooks flourished for many decades after Edison invented his phonograph, and students studied by reading, not listening to Edison's recorded sound cylinders. On the other hand, students, particularly those in STEM (science, technology, engineering, mathematics), now access many of their texts on their laptop computers and do not purchase many hardcover textbooks. These students also learn by audio, listening to podcasts and streaming TED lectures online, so Edison's prediction of sound recordings replacing textbooks turned out to be partially accurate.

Bell, who developed sound amplification methods to help deaf people hear, was firmly against sign language, which is widely accepted today, and insisted that vocalization was the best way for the hearing-impaired to communicate with the hearing world. Ironically, Thomas Edison became increasingly deaf as he got older.

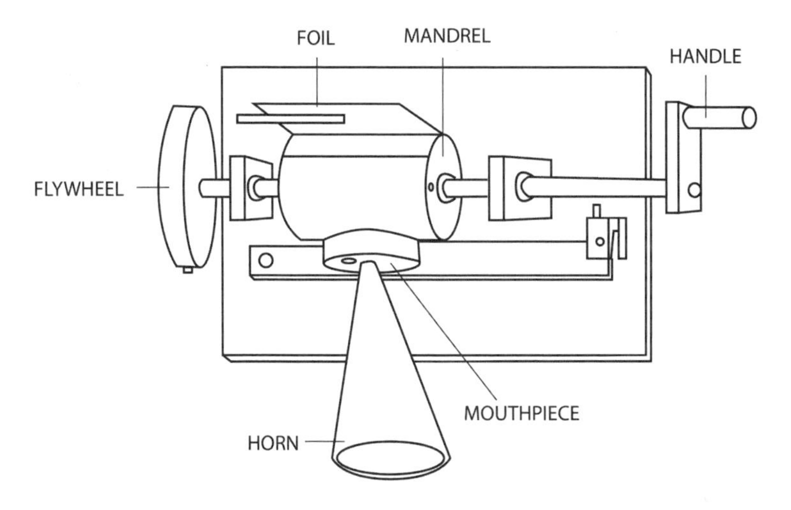

Fig. 8-8. Edison's phonograph.

Edison also launched the motion picture industry and in 1892 he built the world's first movie studio. Although Edison held the patent on the first movie projector, which he called the "kinetescopic camera," his employee William Dickson did most of the technical work. The kinetescope was similar in shape and appearance to one of Edison's sound cylinders, except it was made of silver and coated with an emulsion. The kinetescope cylinder had 42,000 tiny photographic images, each 1/3W2 inch wide, mounted spirally. The movie in the frames of the prototype kinetescope could only be viewed with a binocular eyepiece. For projection on a larger screen, Edison and Dickson transferred the multiple images onto a strip of continuous film. They also developed the first movie camera, called the kinetograph, for filming motion pictures.[18]

Edison's greatest invention is arguably the light bulb, which, powered by the electrical grid, allows us to illuminate the world today. In 1878, Edison conceived the idea of illumination through incandescence: he would pass electric current through some kind of filament, causing the filament to glow with heat and give off light.[19]

On October 21, 1879, after more than 3,000 experiments, many of them testing different filaments without success, he electrified a carbonized cotton thread in a vacuum globe and watched it glow for 45 hours. When a reporter asked Edison how it felt to have so many experiments fail, Edison replied, "The experiments did not fail. They taught me 3,000 ways *not* to make a light bulb!"

Before the light bulb, lighting systems gave illumination primarily through combustion: a rod, filament, wick, or other part of the lamp or candle would burn and be consumed as it gave off flame or glow. Edison's breakthrough was that inside the light bulb existed a vacuum. So the filament glowed with energy, but because there was no oxygen inside, it did not burn. This is why light bulb filaments last many orders of magnitude longer than kerosene lamp wicks.

Thomas Edison was elected to the National Academy of Sciences in 1927. Awarded his last patent in 1928, he held 1,093 patents, the most the U.S. Patent and Trademark Office has ever granted one person. Edison's patents included 141 for storage batteries, 389 for electric light and power, 195 for the phonograph, 150 for the telegraph, and 34 for the telephone.[20] His hundreds of electrical patents included many different kinds of electric lights, dynamos, meters, batteries, battery chargers, electric vehicles, electrodes, electrical switches, relays, transmitters, thermal regulators, and other devices.[21]

Edison's last invention was the spirit phone, a telephone designed to communicate with the dead. Edison believed that if Einstein's law that matter and energy can be neither created nor destroyed was right, then when we died, our consciousness was converted into energy, which might coalesce into a packet. The spirit phone attempted to communicate with this packet through another packet, a stream of photons. The spirit phone converted photons into patterns of electrical charge. It was intended to carry a flow of electrons back from the afterlife to our plane of existence.

More specifically, Edison posited that memories are groupings of electrons, which he called "life clusters." These clusters had an existence independent of the physical body after death. The spirit phone would identify the presence of life clusters for deceased individuals and then create a channel of communication through a device known as the photo cell meter. Just as a megaphone increases the volume of sound, the spirit phone would increase the volume of the electronic registration of the phenomenon that it detected.

Edison was by far not the only person who believed communication with the dead could be achieved through technological or other means. Tesla also worked on his own version of a spirit phone. In the 1941 movie *The Devil Commands*, Boris Karloff plays a scientist who constructs a machine that uses cadavers as human radio tubes for amplifying the incoming spirit signal. Electronic voice phenomenon (EVP) is the term used today for spirit communication achieved through electrical means and was the subject of a 2005 motion picture, *White Noise*, starring Michael Keaton. Both Sir Arthur Conan Doyle, creator of Sherlock Holmes, and magician Harry Houdini pursued communication with souls of the departed.

Though he held more patents than any other individual during his lifetime, Thomas Edison was, as many men of science are, humble in his awe of the natural world, the universe, and their secrets. Expressing the enormous challenge humanity faced in unraveling these secrets, Edison famously said: "We don't know one millionth of one percent about anything."

Edison was somewhat cantankerous. He was a practitioner of trial-and-error research, not much for theory, and made limited use of mathematics. Although the relationship between Steinmetz and Edison began as cordial, it evolved into a bitter war of ideas—specifically, disparate visions of the best way to bring the wonders of electricity to the mass millions of Americans who would eventually use it to light their homes and power the machines and appliances that would make their lives so much easier.

Unlike the socialist Steinmetz, Edison was a dedicated capitalist who ruled his company with an iron fist. When several factory laborers formed a union to gain negotiating power with their boss, Edison responded by building 30 machines to automate their work. The new factory automation made the laborers obsolete; Edison promptly fired them and said, "The union went out. It has been out ever since."[22]

Interestingly, though a socialist, Steinmetz, too, was not a fan of labor unions. He believed the upward mobility of the average American worker made unions unnecessary, as government and big corporations would act in the best interest of blue-collar laborers. Steinmetz also felt unions were counterproductive, because they seemed to want to stifle the business growth and profits that made America prosperous for all.[23]

Between Edison's light bulbs and Steinmetz's electrical grid, homes throughout American became brighter and lighter. Instead of shadowy darkness illuminated only partially by flickering candles and flames in a fireplace, the interior of a house with electric lamps switched on at night could be as bright as day.

There is some debate on how this new nighttime illumination enabled by electrification has affected people's sleep patterns. Some sources say that in the darker nights before electric lights, people went to bed earlier and slept longer. A study in the *Journal of Clinical Sleep Medicine* concludes: "The absence of modern living conditions is associated with an earlier sleep phase and prolonged sleep duration."[24] An article in the *Harvard Health Letter* reports, "Light throws the body's biological clock out of whack [and] sleep suffers."[25]

Edison himself was highly critical of both sleep and sloth, and he believed the latter was a result of too much of the former. He advocated hanging light bulbs throughout rooms, offices, labs, and factory floors to keep people awake and alert. He believed the brightness of his electric light bulbs caused people to sleep less, which in turn increased their energy and productivity and sharpened their minds. Edison claimed he slept only four hours a night, often catnapping in his office or lab. His rival, Nicola Tesla, went to bed at 5 a.m. and got up at 10 a.m., spending five hours lying down—but Tesla claimed he slept for only two hours and just rested in quiet contemplation the other three hours.[26]

As he got older, Edison developed digestive difficulties. Toward the end of his life, his main sustenance consisted of crackers crumbled up in bowls of milk. His reasoning was that if babies could easily digest milk, then an adult with compromised gastrointestinal function should be able to do likewise. Tesla too switched to a diet consisting mainly of milk in his old age. Today, nutritionists know that about three out of four people are genetically incapable of properly digesting milk and other dairy, a condition known as lactose intolerance.

A fictional version of Edison is featured in the 1898 novel *Edison's Conquest of Mars* by Garrett P. Serviss.[27] In the novel, Edison builds 100 spaceships powered by electrical attraction and repulsion. He also invents a ray gun that causes the molecules in any object to vibrate so rapidly that it disintegrates the target.[28]

9

The Bohemian Scientist

Although a hardworking and accomplished scientist and engineer, Steinmetz had many interests and gained a reputation as an eccentric and a bohemian. Those pursuits included gardening with exotic plants, a home zoo on his large estate, hiking, canoeing, bicycling, photography, philosophy, and politics.

While Steinmetz, unlike Edison, never publicly criticized sleep or those who slept a lot, he slept only when he had to, because there were too many things to do and not enough time to do even a fraction of them. He was an active outdoorsman and a lover of nature, and had a plethora of physical and intellectual interests. His mind was ever restless and curious, always thinking and producing.

At the same time, he doted on his three adopted grandchildren and spent long hours with them canoeing, hiking, playing and telling them stories. It was not a duty, burden, or responsibility, this care of younger children. He had always loved children, and he loved his grandchildren more than anything else in the world.

Yet his work was a close second. He routinely worked 12-hour days, and if he found a technical problem particularly compelling, he would continue to work late into the night, eating only when he was hungry and sleeping only when so tired he could not keep working. "To succeed is to make a living at work which interests you," said Steinmetz. "The wise man learns to live. The shrewd man learns to make money. But the man who has learned to live is the happier of the two. Because his work interests him, it is not work at all."[1]

As a socialist, he felt sympathy for the average working man who labored only to pay his bills and did not like his job. He considered himself and others who loved what they did for a living among the fortunate few, observing that if you do what you love for work, you are really not "working" at all. He said just as artists express themselves by painting pictures or composing symphonies or sonnets, engineering was his way of expressing who he was and fulfilling his highest purpose.

Yet unlike, say, Henry Ford, who was singularly focused on business, or Noel Coward, who famously said "work is more fun than fun," Steinmetz filled every waking hour with activities he found interesting, enjoyable, and pleasurable, encompassing work, leisure, and family.

Writer

Steinmetz wrote about 200 published article and 13 books, many of which are about alternating current and are considered the authoritative texts in the field. In his younger days, before joining General Electric, he wrote articles and a couple of books on other topics, including philosophy and astronomy.

Professor Harris Ryan, president of the American Institute of Electrical Engineers, said, "Dr. Steinmetz assisted his brother engineers by an untold degree by his books, papers, and discussions."[2] At GE, when younger engineers were stuck on a problem, they said, "Let's take it to the Supreme Court." The "Supreme Court" was what they called Steinmetz, because he had the wisdom and knowledge to solve the problem and settle any disagreements on technical issues.

Many people pleaded with the great man while he was alive to write his autobiography. However, he didn't think it was important and said he did not have the time. Nonetheless, he would always cooperate with journalists and publicists who had been given the assignment of writing stories about his work. Although Steinmetz was a facile writer and prolific author, writing was not at the core of his being. That, he said, was a role fulfilled by mathematics and engineering, his truest and highest creative outlets for self-expression.

Most people who write books are avid readers, and Steinmetz was no exception. His favorite books included Robert Louis Stevenson's *Treasure Island*, Mark Twain's *Huckleberry Finn*, Rudyard Kipling's *The Jungle Book*, Goethe's *Faust*, and Homer's *Odyssey*.[3]

Union College around the time Steinmetz taught there.

Professor

Union College in Schenectady was founded in 1795.[4] Steinmetz loved teaching and educating others in what he knew, and doing it at Union College was incredibly convenient, as the campus was only a few blocks from the Steinmetz home on Wendell Avenue. From 1902 to about 1912, he taught electrical engineering classes in the morning. Afternoons he worked either at his home laboratory or at the GE plant.

When I received my chemical engineering education in the 1970s, the vast majority of courses engineering students took were engineering, science, mathematics, and a foreign language, preferably (for us chemical engineers) German. We were allowed only a handful of electives in the liberal arts. Back in his day, Steinmetz did not agree with this narrow approach to an engineer's education. He believed that all engineering students should graduate from college with a good general knowledge of English, literature, history, and natural sciences.

As both an author and a university professor, however, Steinmetz wrote about and taught electrical engineering and engineering mathematics at a high level. While his technical books are considered standard reference works in the field, only highly educated specialists with electrical and mathematical backgrounds can understand them.

Likewise, in his classes, his early lectures were complicated and difficult to follow. Yet students found him caring and inspiring, with a genuine concern for helping them master the material. Out of affection they called him "Steiny," and the nickname stuck with him. He was even inducted as an honorary member of the Phi Gamma Delta fraternity on campus.

When he began teaching at the university, students had mixed feelings about attending Steinmetz's lectures. On the one hand, he spoke and wrote equations on the board rapidly, and the material was often too high level for the students to grasp. So many became anxious about keeping up. But Steinmetz made himself available for extra help outside the classroom and was much better at explaining things one on one. Sometimes he admittedly went too far, essentially doing the students' work for them, because it seemed more expeditious than laboriously explaining to the student how to do it himself.

On the other hand, many students loved attending Steinmetz's lectures because they found his enthusiasm, warmth, and love of learning contagious. Also, over time, his teaching skills improved until he finally became the excellent professor he wanted to be.

Atheist

According to a study from the University of Kentucky, roughly one American in four likely does not believe in God.[5] And Steinmetz was among the one-fourth of us who are nonbelievers. Not all scientists are atheists. Many find belief in God compatible with science. One of the pro-God arguments of some scientists is that the universe is so complex, it could not have happened by accident; there must have been an intelligence behind its design. Joseph Priestly, the discoverer of oxygen, said: "When we say there is a God, we mean that there is an intelligent designing cause of what we see in the world around us, and a being who was himself uncaused."[6]

Other scientists say that a belief in God either at best runs completely contrary to science, or at worst is unsupported by science. Biologist Richard Dawkins has stated, "I think the probability of a supernatural creator is very, very low, and I live my life on the assumption that he is not there."[7, 8] Stephen Hawking said, "We are each free to believe what we want and it is my view that the simplest explanation is there is no God. No one created the universe and no one directs our fate."[9]

Steinmetz had a similar view. The methods and point of view of science were so deeply ingrained in Steinmetz that he could not achieve or sustain belief in a supreme being through faith alone. It was impossible for Steinmetz, a firm believer in the methods of science, to believe in God, because the existence of the Lord could not be logically proved.

As his friend Ernest Caldecott, a Unitarian minister at All Souls' Church in Schenectady, New York, said of Steinmetz:

> Religion, he [Steinmetz] declared, *is based on an assumption which cannot be proved.* Time and again religious exponents have been compelled to shift their ground because science has entered and given explanations of phenomena which had previously been interpreted in terms of supernatural intervention.
>
> Steinmetz had stated that "religion deals with the relations of man with the supernatural, with God and immortality, with the soul, our personality and the ego, and its existence or non-existence after death." This he called "the greatest and deepest problem [that] ever confronted mankind."
>
> His views were practically negative on the entire problem; he was simply an unbeliever. Having given serious consideration to the matter he had decided that there was nothing for him to believe. This has placed him before the public as an atheist. The title he did not deny. [I] would put him down as a confirmed agnostic, for an atheist is a person who knows there is no God, and Steinmetz was not of that temperament. Be that as it may, the fact remains that he was a consistent unbeliever.
>
> With that penetrating mind for which he was famous, he saw reason to believe that no almighty goodness ruled this universe. All was according to cause and effect, a process which had wonderfully evolved without personal direction or plan, an amazing series of adaptations, with man but an incident in the whole.
>
> His only complaint against religion or the church was his strong suspicion that the clergy were afraid to tell the truth as they saw it.[10]

Chairman of the board, president, mayor

Steinmetz cared deeply about learning and education, and he had a soft spot for children; his own grandchildren were the light of his life. He was proud of his contribution to higher education but also wanted to help shape young minds prior to college.

So in 1910 Steinmetz ran for and won a seat on the Schenectady Board of Education and was soon appointed its president. He had the support of Dr. George R. Lunn, who was not only mayor of the city but also a socialist like Steinmetz. In fact the Schenectady Common Council had a socialist majority. The council controlled the city school budget and construction projects, and they were supportive of Steinmetz's plans to improve the public school system.

His goals were the construction of more schools, better playgrounds, free textbooks, and lunch programs. The lunch program and free textbooks were soon in place, but getting the new schools and playgrounds built was more of a challenge.

In 1915, at Lunn's urging, Steinmetz ran for city alderman. He won and was promptly appointed council president, a position that he held simultaneously with the school board presidency. With Lunn's blessing, Steinmetz advocated for the development of small parks in multiple city neighborhoods. When Lunn was away from the city on business or holiday, Steinmetz even served as acting mayor of Schenectady.[11]

Ham radio

The lakeside cabin at Camp Mohawk, New York, was equipped with a ham radio. Ham radios are wireless and operated by amateurs as a hobby. Ham radio started in the early 1900s. Steinmetz loved technology, toys, and talk, so it's only natural he was attracted to the hobby. Ham radio continues to be a popular hobby today; over 700,000 ham radio licenses have been issued in the United States by the Federal Communications Commission. Modern ham radio operators can communicate with fellow ham radio operators just about anywhere on the planet.[12]

But for Steinmetz, radio was more than just a hobby. In 1906, Reginald Fessenden, a Canadian professor, became the first person to broadcast a human voice and recorded music over the airwaves. He asked General Electric to come up with a better technology for radio broadcasting than what currently existed. He wanted a device that would limit the radio signal to one frequency for better reception over long distances; he also specified that the unit would provide modulation to be decoded at the receiver into voice and musical sound.[13]

After a couple of years of research, Steinmetz and one of his lab assistants, Ernst Alexanderson, produced an alternator—a generator with AC output—capable of generating waves of 100,000 cycles per second, with a wavelength of 3,000 meters. The alternator was installed in Fessenden's radio station, where it began broadcasting programs in 1906.[14]

Photography

Steinmetz loved his work. Not only the theoretical work of complex mathematics. But also the hands-on mechanical aspects of electrical engineering. So it makes sense that as a longtime amateur photographer he favored subjects such as scientific and technical equipment for his photographs—machines, devices, generators, laboratory experiments—both his own and those of other scientists.

Even more than his inventions, Steinmetz loved to photograph the Haydens, his adopted family, who lived with Steinmetz in his large home for about two decades until Steinmetz passed away in 1923. Many of the scientist's photographs are kept in an archive at the Schenectady County Historical Society in Schenectady, New York, the city where Steinmetz worked for most of his career at General Electric. Schenectady's Museum of Innovation and Science has a collection of 2,000 glass negative photographs taken by Steinmetz.[15]

Not surprisingly, given his aptitude for the technical side of things, Steinmetz delighted in trick photography, creating images using double and triple exposures to superimpose negatives over each other to create special effects, like flying.

Fig. 9-2. Steinmetz Mohawk glider.

One instance of his trick photography concerned a company Steinmetz formed with E. J. Berg, the Mohawk River Aerial Navigation and Transportation Company, which built gliders—not toys but full-size gliders for flying by a human pilot. One model seemed a predecessor of the modern hang glider, which required the pilot to strap into the wings and launch by movement—in the case of the Mohawk glider, by the pilot running down a steep hill on a windy day (fig. 9-2).

Several were built and tested. But none of the gliders or planes flew successfully. So Steinmetz used multiple exposures to make photos in which it looked like the airplanes *were* flying. This was clearly done for fun, not deception: Steinmetz was a man of integrity when it came to the things that mattered most to him, chief among them scientific research, his family, fair play, and the welfare of society.[16] There is no record of any test pilot being harmed in a failed test flight.

Contrary to what the general public seems to believe, many scientists, rather than being narrowly focused on only science and technology, have a wide range of interests, including the arts in general and photography in particular. For instance, Samuel Morse, inventor of the telegraph, was also an early amateur photographer, as well as teacher of photography with the daguerreotype, the first primitive photography method.[17]

Another scientist who was an enthusiastic photographer was Charles Wheatstone. He used photographic equipment of his own invention to take

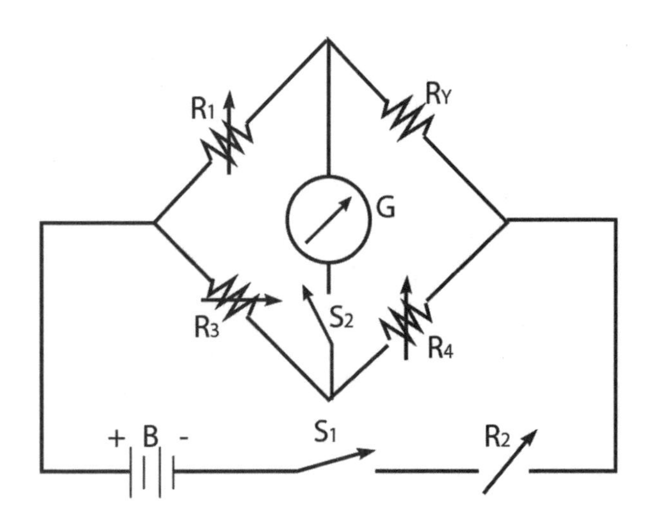

Fig. 9-3. Wheatstone bridge circuit.

pictures of the Wheatstone bridge—an electrical circuit he helped develop, along with Samuel Christie, for measuring electrical resistance (fig. 9-3).

Steinmetz was also one of the earliest amateur photographers, taking pictures with plate cameras for most of his life until he switched to a film camera in 1921. He developed prints from dry glass plate negatives and then later film negatives in his own darkroom. By some accounts, photography may have been the hobby he enjoyed most.

Agriculture

When it came to growing things in the earth, Steinmetz was an enthusiastic amateur gardener but not a gentleman farmer: What he planted on his property were cacti and flowers. He did not grow green vegetables for his own table and in fact seldom ate them; his preferred diet was meat and potatoes.

However, Steinmetz was concerned about food not as a gourmand or for his own nutrition but in terms of world hunger. Although not trained in agricultural science, Steinmetz began experimenting to find a way to raise vast crops of microbes as a source of cheap food. He also had an idea for increasing yields for regular crops, such as corn and wheat. His idea was as follows: Basically, plants depend on nitrogen for growth; in particular, nitrogen is an atom found in the chlorophyll molecule, which is used by plants in photosynthesis to make their food.

However, continual farming depletes the nitrogen content of the soil. Steinmetz's idea was to enrich depleted soil and restore the nitrogen to it. The conventional method of adding nitrogen to soil was with some sort of fertilizer, such as manure or sodium nitrate. But many farmers did not have enough animals to produce sufficient manure to fertilize their fields, and the sodium nitrate was an expensive chemical to buy.

Naturally Steinmetz turned to electricity for a solution. Air is mostly nitrogen, about 78 percent, with 21 percent oxygen, and the remaining 1 percent argon and a few other gases. Lightning converts atmospheric nitrogen into nitrate, a molecule consisting of one nitrogen atom and three oxygen atoms; the nitrate enters the soil with rainfall.[18]

Steinmetz believed, although he did not have a ready answer on how to build it, that a device could be constructed to use electricity to remove pure nitrogen from the atmosphere and return it to the soil. Steinmetz approached Edison, who was even more versatile and varied in his inventing and scientific expertise, and proposed the idea to him. Edison liked the concept and thought it might be workable.

Neither man pursued it beyond the idea stage. But since those days, we have in fact designed and built systems that can distill air into its various gaseous components and extract each element in purified form. Cryogenic plants distill air by gradually lowering the temperature. When the temperature reaches minus 321 degrees Fahrenheit, the freezing point of nitrogen, the nitrogen gas condenses into a thin, clear liquid, which is then separated from the oxygen.

Other nitrogen production systems use membranes to separate nitrogen from air with a technique called pressure swing adsorption (PSA). None of these systems use electricity to remove nitrogen from air directly, but of course they all have compressors, refrigeration coils, and other components that operate on electricity. These systems are used today to generate large volumes of oxygen and nitrogen for both industrial use and health care.

Nitrogen is an odorless, colorless, inert gas, and so a "blanket" of nitrogen gas can be placed over highly reactive and volatile chemicals in a vessel to prevent unwanted combustion. Oxygen is used in medicine, primarily for respiratory therapy. Many hospitals store oxygen in large tanks, where it is transported through a piping system to areas in the building where it is needed. For patients at home, oxygen is supplied as a compressed gas in cylinders. In industry, oxygen is used as a fuel in high-temperature blast furnaces, as pure oxygen combusts at a higher temperature than air.

Astronomy

Steinmetz studied astronomy at the University of Breslau and then wrote a textbook and several articles on astronomy. The textbook was published in Germany and used for several years.[19]

However, Steinmetz appears to have been an "armchair astronomer," more interested in the astrophysics of space and the optics of telescopes than in viewing the cosmos through a telescope. He describes his limited experiments with telescopes, conducted with Joe Hayden, in his book *Radiation, Light, and Illumination.*[20]

Botany and exotic pets

Right next to his huge home on Wendell Avenue in Schenectady, Steinmetz had a large greenhouse which he referred to as his conservatory. It and a few smaller greenhouses on the property were filled with a wide assortment of interesting fauna and flora, including a Gila monster, turtles, crows, and a couple of alligators. He also kept a large aquarium with 22 varieties of tropical fish. In addition, at a house he rented on Liberty Street in Schenectady, before he built his large Wendell Avenue home, Steinmetz kept raccoons, skunks, woodchucks, snakes, and a horse.

In gardening, he preferred odd and unusual plants to pretty flowers, shrubs, and bushes. One writer said of the Steinmetz greenhouse:

> It contained dozens of cacti. They were monstrous plants full of attacking spines and splinters. One misshapen, bulbous cactus was covered with needle-sharp white hair. Another was a snaky, winding horror with clusters of unnatural-appearing fruit growing like ulcers on its body.[21]

This description may strike you as overly dramatic and overblown. It is nothing more than what cacti look like, and plenty of people today like cacti, indoors in the eastern United States where I live and also outdoors in warmer climes. But back in Steinmetz's day, to have an entire greenhouse contain nothing but cacti was a bit of an oddity, compounded by the equally odd appearance of the man who owned it.

Steinmetz explained that his fascination with cacti was that their odd growths, spiny needles, and physiology seemed to be the result of evolution enabling the species to survive and thrive in dry, harsh environments. He believed that the primary characteristic of cacti, which had evolved to ensure the survival of the species, was the thick green trunk. The trunk was green because it contained the plant's chlorophyll, necessary because cacti have no leaves. Evolution had eliminated the leaves, Steinmetz reasoned, because in the strong desert sun, the large evaporating surface of ordinary plant leaves would be a liability. In addition to containing chlorophyll, the cactus trunk and stem held a supply of water needed to survive long dry periods without rain.

He also noticed that cacti have large flowers, better to attract insects capable of fertilizing the plants in the desert where insect life is not abundant. The flowers bloom only at night to let the insects fertilize the pistil with pollen, then close during the day to prevent loss of moisture in the desert heat.[22] He did enjoy the beauty of the cactus flowers, and soon began raising orchids in his greenhouse as well.

Cards

Steinmetz was an active poker player. He and a group of his associates from the plant had a regular poker game almost every Saturday night. He was an average player; while he was great at calculating the odds, he was less adept at bluffing and reading the faces of the other players. Also, not caring much about money, he played for the enjoyment and companionship, and was not focused on how many pots he collected. The group called their weekly poker group "The Society for the Equalization of Engineers' Salaries."

Because of his lack of height and the discomfort caused by his spinal curvature, Steinmetz preferred, when at a poker table, to kneel rather than sit on the chair and then rest his elbows on the table. The stakes were modest. Steinmetz kept track in a ledger of the amount of money won or lost by each player. The men played with chips, not cash on the table.

Cigars

Not only did Steinmetz enjoy a good cigar, but he was rarely photographed without a lit stogie in his hand or clenched between his teeth. As mentioned earlier in the book, Steinmetz was so essential to his company's success that he was the only employee at General Electric who was allowed to smoke in the plant.

Steinmetz said he was not addicted to tobacco and could quit smoking anytime he wanted. When a friend dared him to do so, Steinmetz sure enough gave up cigars for an entire year. As soon as the year was up, he went out, bought a box, and started smoking again, a habit he enjoyed and continued for the rest of his life.

Shorthand

Engineers are problem-solvers, Steinmetz among them. Usually, the problems they work on are technical in nature. But not always. And whether the problem was with electricity or finding a pet alligator of his that had wandered off, Steinmetz liked nothing more than to dive headfirst into what he considered an interesting or important challenge until he found a solution.

One such problem, a dilemma faced by millions of people, both children and adults, including Steinmetz, is to write fast enough to keep up with the thoughts, discussions, interviews, or lectures one wants to take notes on.

The logical solution, before laptop computers, was shorthand. Yet aside from professional secretaries (the shorthand used by secretaries was called stenography), most of us were never taught shorthand, which makes little sense when you consider the importance and difficulty of taking accurate notes, especially when the speaker talks rapidly. The notion of using some sort of shorthand to takes notes quickly is not new. During the fourth century BC, the Greeks created shorthand writing systems that used symbols for letters, words, suffixes, and prefixes.

For his writings, Steinmetz created his own system of shorthand. It was a combination of the Swedish shorthand system along with several others, all customized to his unique code. The Steinmetz shorthand system was based on phonetics. For instance, he wrote the word *height* as *h-i-t.*

Steinmetz claimed that his system of shorthand enabled him to write as fast as he could think. He believed his system to be simple, easy to learn, easy to use, easy to remember, and easy to translate into regular English, and he felt that knowing shorthand was a great advantage to both students and adults alike. The person who takes notes with shorthand can easily keep pace with the teacher or speaker, while the person who writes in longhand often cannot. Also, shorthand is condensed compared with longhand; one page of a Steinmetz shorthand sheet of paper resulted in a three-page manuscript when typed. Steinmetz was somewhat disappointed when his shorthand system was never widely adopted in business or schools.[23]

Shorthand persists today, with the leading shorthand notation being the Gregg system. Gregg was invented by John Robert Gregg (1867–1948) and was first published in 1888. Since then many different versions have appeared, including some for languages other than English. Like the Steinmetz shorthand system, Gregg is phonetic. It represents the sound of speech rather than the correct spelling. So in Gregg, the *f* sounds in the words *form*, *elephant*, and *rough* is written in the same way for each word. Vowels are written as hooks and circles on the consonants.[24]

For most of us today, taking notes via shorthand or longhand with a pad and pen has been replaced by laptop computer. We can type much faster than we can write by hand. When you take notes during a phone interview, you can type on either a desktop or laptop. But when you go to see the speaker at his or her place of business or in the lecture hall, the laptop allows you to keyboard, which was impossible before laptops. Writer Harlan Ellison used to carry a portable manual typewriter when he flew so he could write on airplanes.

Canoe, bicycle, auto

As mentioned, Steinmetz loved to canoe on Lake Mohawk with board and papers in front of him. He liked to think and work on complex mathematical problems related to electrical engineering while enjoying nature. He never learned to sail nor did he buy a bigger boat; the canoe was all he wanted or needed.

Also as discussed earlier, Steinmetz was an avid cyclist. Riding his bicycle was, with the exception of canoeing on the lake, his preferred mode of transportation. With his short legs, hip joint deformity, and limp, he was not a fast walker. But he made much better time zipping between his house on Wendell Avenue and the GE plant, as well as all around town, on his bike.

As mentioned in chapter 7, Steinmetz owned a car and liked to drive, or at least be driven in it. He bought his car from the Detroit Electric Car Company. His personal car was a luxury model. It was outfitted with cut-glass vases for flowers, plush upholstery, and silk currents with tassels. He also bought a second car, a Stanley Steamer.

He had his Detroit Electric car retrofitted with dual controls, one set for the front seat and a second for the rear. Often he would sit up front at the wheel and appear to be driving, while in actuality one of his lab boys was sitting in rear doing the steering.

Socialism

Throughout his life, Steinmetz embraced socialism. He joined the Social Democratic Party as a youth in Germany, which at that time was very much under the official ban. When the editor of the socialist paper was arrested and imprisoned, young Steinmetz secretly assumed his duties. He remained a socialist and, toward the end of his life, ran for state office on the socialist ticket.

He also carried his socialist principles to the almost-unheard-of extreme of refusing an outrageously high salary, though he was paid more than enough to live an extremely comfortable lifestyle. However, he was not a communist. He believed that a man who worked 12 hours a day should be paid more money than one who worked only 4 hours a day, the latter representing the short workday that many in the Socialist Party then seemed to view as ideal. The socialist party in the United States was formed in 1894 by union leader Eugene Debs, who had served six months in jail for organizing a railroad strike in Chicago.

For the greater part of his adult working life, Steinmetz was employed by the General Electric Company. He loved working for this giant corporation, and the company made enormous profits from his work. In return, he got to do what he loved while being paid handsomely for it—and also receiving the funding, assistance, and resources to do his science.

Yet General Electric was then and is today a monolithic corporate giant, a stalwart of the capitalist system that socialism, as both a political party and a worldview, disdains and even sees as evil. So it seems at best a logical contradiction and at worst hypocritical that a staunch socialist would be what his fellow party members would call a lackey of the capitalist system.

Fig. 9-4. Eugene Debs, the founder of the U.S. Socialist Party, speaking against the World War I military draft in 1918.

Socialists essentially believe that powerful groups of capitalists, which are primarily large corporations and wealthy people, are dedicated to lining their own pockets and filling their own coffers, all on the bent backs of the hard-laboring working class. In short, the staunch socialist sees capitalism as being set against the interests and needs of the common man. However, Steinmetz intellectually reconciled his continued support of socialist principles with his belief, based on personal experience and observation, that the large corporation in America is an effective form of economic government.

Steinmetz essentially worked as an employee for one small company, E&O, for a few years, and then for a large corporation, GE, for decades. In his 1916 book *America and the New Epoch*, Steinmetz stated that he much preferred working for a large corporation to working for a small one. Here is the core of his argument for the merits of large corporations over small business.

To begin with, a large corporation operates on a carefully considered strategic plan, designed and managed by executives who are trained businesspeople and know what they are doing. By comparison, many small businesses are run by lone-wolf entrepreneurs who fly by the seat of their pants, without much regard for long-range strategic planning. In the corporate structure, a hierarchy of management and a board of directors lend stability to the company, make management decisions more collaboratively, and ensure that one's continued employment is not at the whim of one autocratic company owner who wields total power over the worker.

Steinmetz also pointed out that large corporations keep profit margins modest and instead rely on their enormous volume of business to ensure adequate net profits; GE, the company that once employed Steinmetz, today has multibillion-dollar annual gross revenues. But smaller companies run by solopreneurs can't handle as many orders. So their revenues are modest, and they must therefore focus instead on large margins to make good profits, a key goal of the owner who most likely wants to get rich.

Small-business owners are often thrifty with paying vendors and employees, because every penny is coming straight out of their own pocket. By comparison, the giant corporations almost always have deeper pockets than the small start-ups. Steinmetz realized that it took a big corporation with a big R&D budget to fund his research without applying pressure to make him produce quick results that could generate an immediate and sizable return on investment.

Steinmetz concluded that the single business owner was an inefficient implementation of capitalism in which the workers were often shortchanged with low pay, stingy benefits, and minimal job security at the whim of the owner and his idiosyncrasies and temperament. He felt the best economic solution is the large corporation—a group of well-paid employees, with good job security, working for a management team, a board, and stockholders whose purchase of shares infused capital into the organization.

In *America and the New Epoch*, Steinmetz used the term *individualism* to describe laissez-faire, the free market system in which business can be conducted with minimum of government regulation, privileges, tariffs, and subsidies, if any. The modern system of large, almost monopolistic trusts he called "cooperation."

He believed socialism was in conflict with individualism, which was fueled by the greed of an individual who sought to enrich himself on the labors of wage-slaves whom he controlled. The entrepreneur's primary goal was not the good of humankind but strictly his own profit. Also, as a small fish in a big pond, the entrepreneur is often forced to be a tightwad concerning his expenses when competing with bigger companies that are better funded. Therefore, the solopreneur's suppliers and workers are often poorly compensated.

Steinmetz also argued that large corporations, especially those not run by a single all-powerful owner, are much better for the economic well-being of the country, for the reasons stated earlier. And so he saw this "cooperation" as more aligned with the core underlying socialist goal of favoring the common laborer. He argued that large corporations take better care of the employees, because the senior executives are themselves most often employees, not owners. He posited that if a corporation grew and distributed stock to its workers, then it was capitalism that embodies the most generous and kindest principles of socialism.[25]

10

Fun, Games, and Over

In his last years, Steinmetz went from serious science and practical technology to more fun but gimmicky, trivial, or even self-indulgent experiments, like the miniature lightning storm. Yes, the lightning machine was also an important invention. But building a miniature village and blasting it to bits with man-made lightning had an element of childlike fun as well as extreme showmanship in it.

After decades of hard technical work and accomplishment, Steinmetz felt entitled to play around with science more and enjoyed doing so. In his 1929 book *Loki: The Life of Charles Proteus Steinmetz*, Jonathan Leonard writes (italics mine):

> Toward the end of his life his scientific work had become rather like a boy's playing with machinery. He would read about an experiment in a technical magazine and then go to his laboratory to try it out for himself. *His great work was done; originality had deserted him.* There remained the small child playing with electricity but a small child accustomed to public praise and avid for more of it . . . in his old age he had become very vain and he loved to have people of no scientific knowledge point to him awe-struck.[1]

Daddy Steinmetz

Much of his science "play" was for the amusement of his three grandchildren—Joe, Midge, and Billy—and their friends. Joe, Midge, and Billy Hayden called their father Father and called Steinmetz Daddy. All the neighborhood children also called him Daddy.

When the Hayden children would bring friends to the home, the kids would ask Daddy to show them some magic. Of course, this was not so much sleight of hand or magic tricks out of a magician's repertoire. All of these tricks were based on some sort of chemical and electrical experiment, but one that had as an objective entertainment and fun rather than scientific discovery.

One favorite experiment was simple liquid color change. Pour a clear chemical into a beaker of water, and the water is still clear. Add another colorless chemical, and this time the solution turns crimson red; Gilbert, once the leading maker of home chemistry sets, called this one "turning water into wine." Add yet another chemical and the red liquid turns clear again.

Steinmetz also amused the children with a simple experiment done with lead. Lead has a low melting temperature, so it is easy to liquefy; you can do it in a pot or pan over an ordinary stovetop flame. He used a ladle held over a Bunsen burner. When the lead had turned to liquid, he poured the molten metal from the ladle into a bucket of cold water.

When you do this, the lead quickly solidifies. Pouring it into a mold is done to make a specific object, such as a bullet. But when the liquid lead fell from the ladle into cold water in an open bucket, it swirled around and then solidified into the random shape of the swirl. Steinmetz and the children took delight in saying what the lead creations reminded them of, much like looking at clouds and seeing figures—a balloon, a horse, or a snake—in their shapes. Warning: Do NOT try this at home, as lead is toxic if absorbed into the bloodstream.

Among the most interesting home chemistry activities are experiments that produce flames. In this Steinmetz did not disappoint. He would drop small pieces of sodium into a bucket of cold water. As soon as the sodium hit the water, hydrogen was released, and the gas bubbles burst into colored flame.

Mrs. Hayden did her best to keep the children out of Daddy's laboratory when he was working. But unlike a lot of workaholics who seem to need peace and quiet to do their creative work, Steinmetz seemed willing, even happy, to have his grandchildren visit him, whatever he was doing. Steinmetz almost always put aside his work when one of his grandchildren came to see him in the lab and instantly focused his attention completely on the child.

The laboratory itself was a great attraction, the sort of secret workshop of a modern-day sorcerer. Steinmetz allowed the children to handle and experiment with apparatus, though always safely, of course. They could add liquid to a test tube, hold it over the flame of the Bunsen burner with a test-tube holder or clamp, and watch the liquid begin to bubble and boil. He permitted them to hold and examine beakers, flasks, and graduated cylinders, though not to handle dangerous chemicals. His only rule was that they stay away from the workbenches on which Steinmetz had current experiments in progress.

Another secret lair in the wizard's castle was Steinmetz's private office, where countless shelves were filled to overflowing with curiosities, keepsakes, various odds and ends, and many collections, including stamps, rocks, butterflies, Indian arrowheads, assorted tools, ostrich eggs, gourds, bottles of formaldehyde preserving the bodies of dead lizards, stuffed animals, even a shrunken head.

While many of the collections and individual items were treasures that Steinmetz's boundless intellectual curiosity had caused him to gather and save, he actively encouraged his grandchildren to add their own finds to what was rapidly growing to the proportions of a small private museum. When the kids brought to Daddy found objects for display, Steinmetz labeled them with small museum cards. For instance, one day Joe brought a rusty nail and it was duly displayed with the legend "Nail dug up by Joe at Ft. Ticonderoga."

The attentions lavished by the kindly grandpa Steinmetz on his grandchildren were not limited to science. He would happily engage them in just about any activity of play that amused either them or him. The activities were many and varied, from a hike at the Lake Mohawk camp and a swim in the Veile Creek to a bicycle ride around town or a day trip in the car, or a building project or story time. Steinmetz seemingly never tired of horsing around with the group of rambunctious kids—his grandkids, their friends, and even neighborhood children who knew that Daddy would greet them warmly and include them in the fun.

But the fun wasn't going to last forever. Steinmetz, once the fun and games were over, felt more tired than he did only a few years ago. Although he was still a workhorse, his energy was no longer boundless. Unbeknownst to Joe Hayden, his father's small, frail body was beginning to wind down.

Practical jokes

Steinmetz had a mischievous side to his personality, though it was for fun and not malicious. This side of him he expressed through an endless stream of practical jokes, played mainly on colleagues and others who came to visit him at his home, lab, or office.

His setups were often, for a technical wizard, extremely simple and primitive. One of his favorites was to paint a simple box the same color as the floor and walls, nail it to the entryway to his lab, and then watch as visitors stumbled over it. In reality, this stunt was not very thoughtful, probably unpleasant to the victim, and actually unsafe. But it greatly amused him, and he never tired of it.

Another trick that also showed perhaps an appalling lack of maturity was sawing a leg on a wooden chair in the lab, making it unsteady. He was an extremely intelligent man. Did he truly not realize that falling the wrong way could cause a lot of pain and possibly back injury?

As Steinmetz was the electrical wizard of Schenectady, his other practical jokes naturally involved electricity. For instance, an associate gave Steinmetz an old static electricity generator. Steinmetz would use it to charge himself so that when he shook a visitor's hand, the person received a mild electric shock. He also attached the generator wires to doorknobs to deliver a shock to anyone opening the door.

For Steinmetz, a known cigar aficionado, it was only natural to offer visitors a cigar. Naturally, they were surprised to discover they were lighting up a trick cigar, as the tip quickly exploded with a small but harmless bang.

Steinmetz had another scientific practical joke, based on his work with lighting. He installed a few mercury vapor lamps in his conservatory. These lamps use a high-intensity electric arc that sparks between two electrodes. The powerful electric current travels through a cloud of vaporized mercury sealed within the tube.[2] When the other lighting was switched off, and the mercury lamps were the only illumination, the already hideous alligators, Gila monster, and bizarrely shaped cacti looked even odder.

To get an even bigger laugh for himself, he then placed a full-length mirror in the entryway to the conservatory, and also installed another mercury map there. He would walk with a visitor until the person was standing right in front of the mirror and then switch on the mercury lamp. The person would be staring at a reflection that had green skin and purple lips, looking somewhat like The Joker![3]

The world's electrical future

In 1915, two years before America entered into World War I, Steinmetz was asked by a newspaper reporter if he thought the process of civilization would be set back by the conflict. On the contrary, he saw a glorious future for the world and humankind and believed much of it stemmed from the electrical systems, products, and technologies that Steinmetz, Edison, Tesla, and his other fellow electrical wizards helped to create. He told the reporter:

> It won't be long before, thanks to electricity, men will only work six hours a day, five days a week. International radio broadcasts will be commonplace and millions will hear the finest orchestras and opera companies giving concerts right in their homes. The motion picture and the talking machine will be perfectly synchronized; while buildings and homes will be heated, cooled, and ventilated by electricity.
>
> Much of our cooking will be done on electric stoves equipped with a dial that will automatically start and stop the heating unit, so that housewives will be able to put their dinners in the stove, leave their houses, and come back and find their meals ready to put on the table. The labor of the farmer will be made lighter by electricity, and the power itself will be much cheaper than it is now, as well as available to many thousands who do not as yet enjoy its advantages.
>
> Now all this means that the world needs men who know something about electricity [and] of the creation and control of electric power. It needs all of us and will need more of us every year. The human race will always need electricians. Its very existence will depend on them.
>
> No one can limit the use of electricity. Electricity is energy and therefore can do anything energy can do. Electricity is energy and energy is the basis of civilization—and through it unknown horizons lie before us.[4]

In a separate interview with another reporter for a different article, Steinmetz predicted a glorious future for electricity and humankind:

> Little by little, our developments in science make the world a better place to live in. We call this the age of electricity, but it isn't. The age of electricity hasn't begun. When the age of electricity comes—as it will—electricity will do for everybody all that it can do for anybody. It will do this in addition to doing a multitude of things of which we have not yet dreamed. And it will be a great age.[5]

Over . . .

In his final years, Steinmetz stepped up his travel activities, both for long-overdue vacations with the family and for trips to lecture, accept honors, and confer and hobnob with his fellow electrical wizards.

After a train trip to see the Grand Canyon, he became fatigued. The doctor who came to the Wendell Avenue home to examine Steinmetz delivered some frightening news: the fatigue was not simple exhaustion. Yes, the scientist had overdone it. But the doctor told him that his heart was not in good condition. He advised bed rest and said he would see Steinmetz again in a couple of days.

Steinmetz was so exhausted, he uncharacteristically listened to the doctor's instructions and agreed to rest without physical activity. He spent his time in bed reading popular suspense and adventure novels. He also had the GE plant librarian deliver to him a scientific text on a subject that he and Joe had become interested in of late, the physics of air and atmospheric changes.

The next morning Joe Hayden came in to check on his dad. When Joe asked Steinmetz how he was doing, the scientist replied, "Oh, don't worry; I'm okay," and then added: "You won't be bothered with me much longer."

Joe thought not being bothered meant Steinmetz would soon leave his sickbed to return to his work and active life. He told Steinmetz to lie back and rest; he would send his son Billy upstairs shortly with the old man's breakfast.

When young Billy Hayden entered his grandfather's room, he knew something was wrong. He stepped closer, and it was then he saw that Steinmetz was not breathing.

Billy shouted downstairs to his father. Joe Hayden quickly bounded up the stairs, entered the room, and, to his sorrow, saw that the great man and kindly father he adored, Charles Proteus Steinmetz, had quietly and peacefully passed away at age 58.[6] His physician reported that he had died from "acute dilation of the heart, following a chronic myocarditis of many years standing, which is a weakening of the heart muscles."[7]

Here is the obituary for Charles Steinmetz as it was published in the October 27, 1923, issue of the *New York Tribune*, one day after his death:

Dr. Steinmetz, Electric Genius, Dies Suddenly

Great Mathematician and Engineer Suffered Strain on Vitality on Trip to Pacific Coast Recently

Achievements Colossal—Most Important Inventions Made in General Electric Plant at Schenectady

Schenectady, Oct. 26, 1923—Charles Proteus Steinmetz, one of the world's greatest mathematicians and electricians, died of sudden heart failure this morning at his home in this city. He had recently returned from a trip to the Pacific coast, during which his physical vitality seems to have been overtaxed. For the last two weeks he had been under the care of a physician and nurse, but was regarded by them as doing well and as making progress toward complete recovery.

A little after 8 o'clock this morning his breakfast was taken to him as he lay in his bed. Just as it was placed before him he expired, peacefully and without warning.

He never married. In stature he was almost a dwarf, but he had a massive head and brilliant eyes, commanding the attention of everyone who saw him. He was an almost incessant smoker of cigars made expressly for him, which were very long and very mild, providing he said, a maximum of smoke with a minimum of nicotine.

His special interests in electrical science were magnetics, the symbolic method of alternating current calculations, and transient phenomena; but there was scarcely a detail of any branch of electrical science, of astronomy or of mathematics, with which he was not conversant in masterly fashion. His writings were voluminous.

Wizard's Death Mourned as Grave Loss to Science

Prominent persons throughout the city expressed deep regret yesterday on hearing of the death of Dr. Steinmetz and reviewed his remarkable list of achievements in electricity. Among them were Thomas A. Edison and Dr. F. P. Jewett, vice-president of the Western Electric Company and formerly president of the American Institute of Electrical Engineers. Mr. Edison said:

"The world has lost one of the greatest practical mathematicians and the electrical industry will miss one of its shining lights. I regret very much to learn of the death of Mr. Steinmetz."

Mr. Jewett said:

"In the death of Dr. Steinmetz, the electrical industry, not alone of the United States but of the world at large, loses one of its conspicuous and distinguished members.

"Surmounting physical afflictions which would have justified a quiet life, he brought to the support of a fertile brain and a vivid imagination an almost incredible energy. For years he was a leader in the field of electrical research, particularly in matters pertaining to machine design and the transmission of energy, and his work in this direction has added much to our knowledge of the mathematical tools for solving complex electrical problems.

"As president of the American Institute of Electrical Engineers in 1901-1902, and throughout his lifelong work in its behalf, he did much to bring it to its present high place as one of the greatest professional engineering societies of the world."

Acting Mayor Murray Hulbert said:

"The death of Charles P. Steinmetz is an incalculable loss not only to the people of this state and nation, but of the world. His wonderful intellect, devoted to the scientific study and development of electrical power, had already produced such beneficent results as to warrant a belief in the public mind that had the life of this electrical wizard been spared a decade longer the results in the development of economical light, heat, and power would have been well-nigh inestimable. Devoted as he was to industrial scientific research and beneficial discoveries to mankind, his loss is universal. His life was the arduous greatness of things done and the hope of still greater benefits to mankind."

Arthur Williams, general commercial manager of the New York Edison Company and president of the Electrical Board of Trade of New York, said:

"In the death of Dr. Steinmetz the scientific world loses one of the greatest minds it has ever known. His achievements and contributions to science and industry the world over, accomplished under the most trying physical conditions, will remain always an inspiration to those who are acquainted with his life's work. The great things he accomplished will always mean much for the betterment of the civilized world."

The body of Dr. Steinmetz will lie in state at his home here tomorrow. Burial will be on Monday afternoon, after private funeral services, at Vale Cemetery in a plot which the inventor acquired several years ago. The Rev. Ernest T. Caldecott, pastor of All Souls' Unitarian Church, will officiate at the services, assisted by the Rev. Dr. A. W. Clark.

Though the diminutive body of Steinmetz has surely returned to dust and bone, the ideas that sprung from this intellectual giant's mind, expressed in both his writings and his inventions that now bring power and light to humankind around the world, will shine bright and live forever, or at least until the end of civilization as we know it.

Timeline of Steinmetz's Life

1865 Charles Proteus Steinmetz born in Germany.

1889 Leaves Europe and immigrates to the United States; hired as a draftsman by Eickemeyer and Osterheld (E&O) in Yonkers, New York.

1890 *Electrical Engineer* publishes Steinmetz's paper "Note on the Law of Hysteresis."

1892 General Electric Company begins operations under its own name.

1893 Hired as chief consulting engineer of the General Electric Company.

1894 Steinmetz becomes U.S. citizen.

1895 Steinmetz patents a distribution system for AC power.

1897 Experiments with a unique single-phase AC power transmission system at the Mechanicville Power Station.

1901 Elected president of the American Institute of Electrical Engineers.

1902 Appointed professor of electrophysics at Union College.

1902 Awarded honorary degree from Harvard.

1903 Awarded honorary PhD from Union College.

1905 Adopts adult lab assistant Joseph Hayden as his son.

1907 Designs a transformer capable of stepping up power to an unprecedented 220,000 volts.

1909 Patents his design for an electric arc street lamp.

1912 Appointed president of the Schenectady Board of Education.

1915	Elected to the Schenectady Common Council on the socialist ticket.
1920	Steinmetz works out flow of electricity in lightning bolts after the mirror in his summer cottage is struck and shattered by one.
1920	Formation of Steinmetz Electric Motor Company to manufacture electric cars and trucks; first successful road test of truck in 1922.
1922	Dramatic public demonstration shows the effects of lightning by destroying a model village with artificial lightning.
1923	Steinmetz dies at age 58.

Selected Books, Papers, and Lectures by Charles Steinmetz

America and the New Epoch, Charles P. Steinmetz. (New York: Harper & Brothers, 1916).

"America's Energy Supply," Charles P. Steinmetz, AIEE Trans., XXXVII (2):985–1014, Jul. 18, 1918.

"The Alternating Current Induction Motor," Charles P. Steinmetz, AIEE Trans., XIV (1): 183–217, 1897.

"The Natural Period of a Transmission Line and the Frequency of Lightning Discharge Therefrom," Charles P. Steinmetz, The Electrical World, August 27, 1899, 203–205.

"The Alternating-Current Railway Motor," Charles P. Steinmetz, AIEE Trans., XXIII: 9–25, Jan. 1904.

"Cable Charge and Discharge," Charles P. Steinmetz, AIEE Trans., XLII: 577–592, Jan. 1923.

"Charles Steinmetz: Scientist and Socialist." Three American Radicals: John Swinton, Charles P. Steinmetz, and William Dean Howells. Sender Garlin. (Boulder, CO: Westview Press, 1976.

"Collectivism and Charles Steinmetz", James B. Gilbert, Business History Review, vol. 48, no. 4 (Winter 1974), 520–540.

"Condenser Discharges Through a General Gas Circuit," Charles P. Steinmetz, AIEE Trans., XLI: 63–76, Jan. 1922.

"Disruptive Strength with Transient Voltages," Charles P. Steinmetz, with Hayden, Joseph L. R. principal author, AIEE Trans., XXIX (2): 1125–1158, May 10, 1910.

Elementary Lectures on Electric Discharges, Waves and Impulses, and Other Transients. Charles P. Steinmetz. (New York: McGraw-Hill, 1911.)

"Electrical Engineering Education," Charles P. Steinmetz, AIEE Trans., XXVII (1): 79–85, Jan. 1908.

Engineering Mathematics: A Series of Lectures Delivered at Union College. Charles P. Steinmetz. (New York: McGraw-Hill, 1911).

"Essay on Science and Religion." Charles P. Steinmetz. In A Book of Exposition, Homer Heath Nugent, ed. (New York; Harcourt, Brace, 1922).

Four Lectures on Relativity and Space. Charles P. Steinmetz. (New York: McGraw-Hill, 1923).

"Frequency Conversion by Third Class Conductor and Mechanism of the Arcing Ground and Other Cumulative Surges," Charles P. Steinmetz, AIEE Trans., XLII: 470–477, Jan. 1923.

"Future of Electricity," Charles P. Steinmetz, Transcript of lecture to the New York Electrical Trade School, 1908.

"The General Equations of the Electric Circuit," Charles P. Steinmetz, AIEE, Trans., XXVII (2):1231–1305, Jun. 8, 1908.

"The General Equations of the Electric Circuit-III," Charles P. Steinmetz, AIEE Trans., XXXVIII (1):191–260, Jan. 1919.

General Lectures on Electrical Engineering. Charles P. Steinmetz. Joseph Le Roy Hayden, ed. (Schenectady, NY: Robson & Adee, 1908).

"High-Voltage Insulation," Charles P. Steinmetz, with Hayden, J. L. R. principal author, AIEE Trans., XLII:1029–1042, Jan. 1923

"Instability of Electric Circuits," Charles P. Steinmetz, AIEE Trans., XXXII (2):2005–2021, May 13, 1913

"Lightning Phenomena in Electric Circuits," Charles P. Steinmetz, AIEE Trans., XXVI (1):401–423, Jan. 1907

"Mechanical Forces in Magnetic Fields," Charles P. Steinmetz, AIEE Trans., XXX (1):367–385, Jan. 1911

"On the Law of Hysteresis," Charles P. Steinmetz, AIEE Trans., IX: 3–64, 1892; Proc. of the IEEE, 72(2):197–221, doi: 10.1109/PROC.1984.12842

"Overdamped Condenser Oscillations," Charles P. Steinmetz, AIEE Trans., XLIII: 126–130, Jan. 1924

"The Oxide Film Lightning Arrester," Charles P. Steinmetz, AIEE Trans., XXXVII (2):871–880, Jul. 18, 1918

"Power Control and Stability of Electric Generating Stations," Charles P. Steinmetz, AIEE Trans., XXXIX (2):1215–1287, Jul. 20, 1920

"Primary Standard of Light," Charles P. Steinmetz, AIEE Trans., XXVII (2):1319–1324, Jun. 8, 1908

"Prime Movers," Charles P. Steinmetz, AIEE Trans., XXVIII (1):63–84, Jan. 1909

Radiation, Light and Illumination. Charles P. Steinmetz. (New York: McGraw-Hill, 1918).

"Recollections of Steinmetz: A Visit to the Workshops of Dr. Charles Proteus Steinmetz, "Emil J. Remscheid & Virginia Remscheid Charves, General Electric Company, Research and Development, 1977.

"Recording Devices," Charles P. Steinmetz, AIEE Trans., XXXIII (1):283–292, Jan. 1914

"Some Problems of High-Voltage Transmissions," Charles P. Steinmetz, AIEE Trans., XXXI (1):167–173, Jan. 1912

"Speed Regulation of Prime Movers and Parallel Operation of Alternators," Charles P. Steinmetz, AIEE Trans., XVIII: 741–744, Jan. 1901

Theory and Calculation of Transient Electric Phenomena and Oscillations. Charles P. Steinmetz. (New York: McGraw-Hill, 1909).

"Theory of the General Alternating Current Transformer," Charles P. Steinmetz, AIEE Trans., XII: 245–256, Jan. 1895

Theory and Calculation of Alternating Current Phenomena. Charles P. Steinmetz, with the assistance of Ernst J. Berg. (New York: Electrical World and Engineer, 1900). [Information from this book has been reprinted in many subsequent engineering texts.]

Theoretical Elements of Electrical Engineering. Charles P. Steinmetz. (New York: McGraw-Hill, 1902).

Theory and Calculation of Electric Apparatus. Charles P. Steinmetz. (New York: McGraw-Hill, 1917).

"Outline of Theory of Impulse Currents," Charles P. Steinmetz, AIEE Trans., XXXV (1):1–31, Jan. 1916

"The Value of the Classics in Engineering Education," Charles P. Steinmetz, AIEE, Trans. XXVIII (2):1103–1106, Jun. 9, 1909

Key People in the Development of Electricity

Benjamin Franklin (1706–1790), American

Franklin famously flew a kite in a thunderstorm. The kite picked up electrostatic charges from the electrified atmosphere, and the charges were conducted down the kite string into an electrical storage device called a Leyden charge. However, he was by no means the discoverer of electricity as he is sometimes erroneously thought to be or even the first to attract electricity by flying a kite in a storm. Franklin was a successful scientist, statesman, entrepreneur, and printer. He is credited with originating the idea of free lending libraries in the United States. During the Revolutionary War, he traveled to France and convinced the French to provide naval support to the colonial troops against Britain.

In addition, his kite experiment had shown Franklin that electricity discharges more readily when brought close to a pointed object. Franklin built pointed rods and placed them above the roofs of buildings, with wires leading to the ground, inventing the first lightning rods; by 1782, there were 400 such lightning rods installed throughout Philadelphia.

Charles-Augustin de Coulomb (1736–1806), French

In 1785, Coulomb demonstrated that the force of electrical attraction of repulsion is proportional to the product of the charges on two charged spheres, as well as inversely proportional to the square of the distance between the centers of the spheres. The unit for electric charge, the coulomb, is named after him. Joseph Priestly and Henry Cavendish had reached the same conclusion years earlier, but Coulomb published his experimental results while Cavendish did not, and Priestly reached his conclusion by conjecture rather than direct evidence.

James Watt (1736–1819), Scottish

Watt developed an improved steam engine. To quantify the power his steam engine produced, he compared it with the work performed by a strong hose, which could raise a 150-pound weight nearly 4 feet in one second. He defined one "horsepower" as 550 foot-pounds per second. In his honor, the unit of power in the metric system is the watt; one horsepower is equal to 746 watts.

Count Alessandro Volta (1745–1827), Italian

Volta built a battery made up of a pile or stack of alternating materials consisting of anodes and cathodes with an electrolyte-soaked absorbent material between each metal plate. It is called a Voltaic pile. In his honor, the unit of the electromotive force, the driving force that moves electric current, is called a volt.

André M. Ampère (1775–1836), French

Ampère was the first to apply advanced mathematics to the analysis of electricity and magnetism. The quantity of electric current passing a given point in a given time is named the ampere, or amp, in his honor.

Hans C. Oersted (1777–1851), Danish

While teaching a university class, Oersted brought a compass needle near a wire that was conducting an electric current. When he saw the compass needle move as the wire was brought near it, he continued to experiment with it, concluding that this action demonstrated a connection between electricity and magnetism.

Carl F. Gauss (1777–1855), German

Gauss, a mathematician, correctly calculated the position of the earth's magnetic poles based on collected geomagnetic data. For his work in magnetism, the unit of measure for magnetic flux density, the gauss, was named after him. The earth's two magnetic poles, north and south, are not quite on

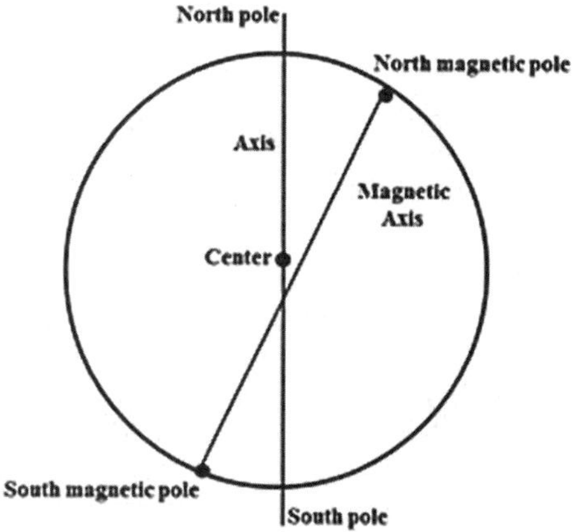

Fig. A-1. The earth's magnetic poles.

opposite sides of the earth, so the line connecting them, called the earth's magnetic axis, comes close to but does not pass through the exact center of the planet. The flowing of liquid metal in the earth's outer core generates electric currents; the rotation of Earth on its axis causes these electric currents to form our planet's magnetic field, which prevents our atmosphere from being blown away by solar winds.

Georg S. Ohm (1789–1854), German

A German physicist, Ohm figured out that the amount of electrical current transmitted is inversely proportional to the length and directly proportional to the cross-sectional area of a given wire. His most important discovery was Ohm's law, which says that the flow of current through a conductor is directly proportional to the potential difference and inversely proportional to the resistance.*

These relationships are expressed in the formula Ohm's law, R=E/I. Here R is resistance, E is voltage, and I is current intensity. The ohm, a unit of measure for resistance, is named for Georg Ohm.

Michael Faraday (1791–1867), English

In 1831, Faraday made what is arguably the most important contribution to the modern system of power generation, electrical inductance, which is that moving a coil of metal wire through a magnetic field produces electricity. Faraday made many other important scientific discoveries, including Faraday's laws of electrolysis. Electrolysis is a process of liberating new metals by passing an electric current through molten compounds of those metals. Faraday discovered that the mass of a substance liberated at an electrode during electrolysis is proportional to the quantity of electricity driven through the solution.

Joseph Henry (1797–1878), American

Henry actually discovered electrical inductance before Faraday, but Faraday published his results first and is therefore given the credit. However, Henry is credited for publishing in 1831 the first paper to propose an electric motor, in which electricity turns a wheel to produce mechanical force. Faraday's electrical inductance is the basis of all generators, while Henry's motors are a primary means of harnessing electricity to make all sorts of

* Electrical potential is the amount of work needed to move a unit positive charge from a reference point to a specific point inside an electromagnetic field without producing any acceleration.

electrical appliances—from refrigerators and air conditioners to vacuum cleaners and electric shavers—work.

James P. Joule (1818–1889), English

Joule was the first to accurately determine the mechanical equivalent of heat, which is that 41.8 million ergs of work always produce one calorie of heat. Under normal circumstances, energy cannot be created or destroyed, yet in machinery, some mechanical energy is lost through air resistance and friction, which generate heat. Joule's mechanical equivalent of heat showed that the amount of loss of mechanical energy is exactly equal to the energy gained in heat—and so total energy is indeed conserved.

James C. Maxwell (1831–1879), Scottish

In his 1865 book *A Dynamic Theory of Electrical Field*, Maxwell was among the first to work out the rigorous mathematics that expressed the varied aspects of electricity and magnetism, including the mathematics of how Faraday's magnetic conduction coil generated electricity (Faraday was largely self-taught in science and did not know higher-level mathematics). Maxwell also stated that light was electromagnetic radiation, and he predicted the existence of radio waves. Steinmetz's analysis carried Maxwell's work further and applied it more specifically to alternating current and electrical circuits.

George Westinghouse (1846–1914), American

As an industrialist, Westinghouse manufactured the alternating current equipment designed by Nikola Tesla. His company, Westinghouse, developed the electrical generator powered by Niagara Falls.

Thomas A. Edison (1847–1931), American

No American inventor had as many different patents for different inventions in different fields than Thomas Edison. He invented the light bulb, which became a primary motivation for scientists and engineers to develop the U.S. electrical power distribution grid. But he believed the grid should carry direct current rather than alternating current, and in this lost out to Steinmetz, Tesla, and Westinghouse. He also invented motion pictures and the phonograph.

Nikola Tesla (1856–1943), Serbian/American

Nikola Tesla was, along with Charles Steinmetz, the premier developer of the electrical grid for the transmission and distribution of alternating current. He had almost as many inventions to his credit as his one-time colleague and later rival Thomas Edison; these included a remote-controlled torpedo, wireless power transmission, and the radio, though Marconi, not Tesla, took credit for the invention of the latter.

Heinrich R. Hertz (1857–1894), German

Among his many discoveries, Hertz was the first to verify Maxwell's equations and experimentally demonstrate the existence of electromagnetic waves. The unit of wavelength frequency, the hertz, is named after him.

Hertz developed what was probably the first equipment capable of generating electromagnetic waves. He demonstrated that electromagnetic waves could be reflected and refracted just like light and radiant heat, which he showed were in fact also electromagnetic waves.

Charles P. Steinmetz (1865–1923), German/American

Steinmetz was a principle architect and designer of today's electrical grid for distribution of alternating current. He is credited with working out a rigorous mathematical explanation of the behavior of alternating current and designing circuitry to transmit electricity more efficiently over greater distances.

Notes

Introduction

1. Kline, Ronald,. *Steinmetz: Engineer and Socialist* (Baltimore: Johns Hopkins University Press, 1992), 314.
2. Smithsonian.com.
3. Martin, Vaughn, "Charles Steinmetz: The Father of Electrical Engineering," *Nuts Volts*, 2009, 1.
4. Electric Power Research Institute, *Pictorial History of the Electricity Industry.* 23, 2015.
5. http://electronlab.org/articles/.
6. http://www.eham.net/articles/9517.

Chapter 1: Coming to America

1. http://www.history.com/this-day-in-history/statue-of-liberty-arrives-in-new-york-harbor.
2. http://behindthescenes.nyhistory.org/castle-garden-where-immigrants-first-came-to-america.
3. Miller, Floyd, *The Man Who Tamed Lightning* (New York: Scholastic, 1966), 9.
4. https://en.wikipedia.org/wiki/History_of_laws_concerning_immigration_and_naturalization_in_the_United_States.
5. Hart, Larry, *Schenectady's Golden Era: 1880–1930* (Schenectady, NY: Old Dorp Books, 1974), 17.
6. Steinmetz, Charles Proteus, *Four Lectures on Relativity and Space* (New York: McGraw-Hill, 1923), 23.
7. https://www.britannica.com/biography/Charles-Proteus-Steinmetz.
8. https://www.marxists.org/archive/morris/works/1885/manifst2.htm.
9. http://web.mnstate.edu/shoptaug/AntiFrames.htm.
10. Kline, 29.

11. Ibid, 33.
12. Hammond, John, *Men and Volts: The Story of General Electric* (Philadelphia: J. B. Lippincott, 1941), 303.
13. Steinmetz, Charles Proteus, *America and the New Epoch* (New York: Harper & Brothers, 1916), 172–173.
14. http://www.gestaltreality.com/energy-synthesis/eric-dollard/first-posts-by-dollard-e-p-n6kph/.
15. https://sciencing.com/metals-make-good-conductors-electricity-8115694.html.
16. http://www.bbc.co.uk/schools/gcsebitesize/science/add_ocr_gateway/periodic_table/metalsrev1.shtml.
17. https://www.copper.org/applications/electrical/building/wire_systems.html.

Chapter 2: The Evolution of the Electric World

1. https://economictimes.indiatimes.com/indias-looming-power-crisis/articleshow/51051903.cms.
2. https://www.greenbiz.com/article/fight-over-how-power-developing-world.
3. http://www.cnn.com/2016/04/01/africa/africa-state-of-electricity-feat/index.html.
4. https://theconversation.com/what-lies-behind-africas-lack-of-access-and-unreliable-power-supplies-56521.
5. https://www.washingtonpost.com/graphics/world/world-without-power/.
6. http://insideenergy.org/2014/08/18/power-outages-on-the-rise-across-the-u-s/.
7. http://abc7.com/news/94k-left-without-power-in-san-fernando-valley-after-ladwp-explosion-fire/2197507/.
8. *Off the Grid News*, July 27, 2017.
9. Ibid., July 19, 2017.
10. *Power Markets Today*, March 28, 2018.
11. *Off the Grid News*, November 15, 2017.
12. Roberts, Adam, *The History of Science Fiction* (London: Palgrave Macmillan, 2016), 318.
13. *Off the Grid News*, May 17, 2017.
14. http://www.marketsandmarkets.com/PressReleases/generator-sales.asp.
15. http://www.energycentral.com/c/iu/denvers-peÃ±a-station-next-project-proving-ground-smart-city-concepts.
16. Lowe, Derek, *The Chemistry Book* (New York: Sterling, 2016), 11.
17. http://www.edisontechcenter.org/batteries.html.
18. http://www.qrg.northwestern.edu/projects/vss/docs/power/2-how-do-batteries-work.html.

19. https://engineering.mit.edu/engage/ask-an-engineer/how-does-a-battery-work/.
20. http://www.cisco.com/c/en/us/training-events/esd-training-program/how-much-static.html.
21. http://farside.ph.utexas.edu/teaching/302l/lectures/node12.html.
22. http://scienceline.ucsb.edu/getkey.php?key@26.
23. https://micro.magnet.fsu.edu/electromag/java/faraday2/.
24. Nagpurwala, Q. H., "Design of Steam Turbines" (Lecture, M. S. Ramaiah School of Advance Studies, Bangalore).
25. Solomon, Robert, *The Little Book of Mathematical Principles* (Grantham, Lincolnshire: IMM Lifestyle Books, 2016), 129.
26. "Gravitational Waves Shows Us Neutron Stars Colliding" (Lecture, TED Science Worth Knowing, December 21, 2017).
27. http://164.100.133.129:81/eCONTENT/Uploads/13-PT12-Steam turbines-New0810 63 [Compatibility Mode].pdf.
28. https://energy.gov/fe/how-gas-turbine-power-plants-work.
29. "Fire at Natural Gas Power Plant," *Industrial Equipment News*, November 30, 2017.
30. *Power Markets Today*, December 9, 2017.
31. https://www.ien.com/operations/news/20999985/coal-plant-shuttered-in-wi.
32. *The Utility Business Customer Survey on Energy Management* (Spokane, WA: Ecova, 2017).
33. *Smart Grid Today*, February 14, 2018.
34. Solomon, Dan, "The Future's So Bright," *Smithsonian*, April 2018, 68.
35. Asimov, Isaac, *Opus 100* (Boston: Houghton Mifflin, 1969), 141.
36. Solomon, Dan, 71.
37. *Smart Grid Today*, July 19, 2017.
38. Ibid., April 2, 2018.
39. *Oil and Energy Investor*, May 23, 2018.
40. https://oilandenergyinvestor.com/2018/05/how-to-lock-down-legacy-wealth-from-solars-golden-era.
41. https://blogs.scientificamerican.com/plugged-in/wind-energy-is-one-of-the-cheapest-sources-of-electricity-and-its-getting-cheaper/.
42. https://www.instituteforenergyresearch.org/analysis/the-hidden-costs-of-wind-power/.
43. *Generation Network*, April 17, 2018.
44. *The Capitalist*, March 30, 2018.
45. Miller, 4.
46. Hall, Shannon, "Hot Rocks," *Scientific American*, April 20, 2018, 17.
47. "Duke Energy Ties Pig Poop Gas into Power Plant," *Industrial Equipment News*, March 29, 2018.

48. *Industrial Equipment News*, April 6, 2018.

49. Solomon, Dan, 74.

50. Meeusen, E. J., *Fundamentals of Electricity: Volume 2: Alternating Current* (Boston: Addison-Wesley, 1966), 8.

51. Golder, David, *The Astounding Illustrated History of Science Fiction* (London: Flame Tree, 2017), 31.

52. https://www.eia.gov/tools/faqs/faq.php?id 7&t=3.

53. http://www.hydro.org/policy/faq/#882.

54. http://web.archive.org/web/20170301080617/https:/energy.gov/articles/top-9-things-you-didnt-know-about-americas-power-grid.

55. *Smart Grid Today*, July 25, 2017.

56. *Energy: State of the Distributed Grid 2017* (Bethesda, MD: Lockheed Martin, 2018), 3.

57. Furgang, Kathy, *Zoom in on Power Grids* (New York: Enslow, 2017).

Chapter 3: The Mathematics of Electricity

1. Kleitman, Daniel, *Calculus for Beginners and Artists*. Spring 2005, MIT OpenCourseWare, https://ocw.mit.edu.

2. AIChE Engage

3. Ibid.

4. Kline, 305.

5. Lavine, Sigmund A., *Steinmetz: Maker of Lightning* (New York: Dodd, Mead, 1955), 169.

6. Stein, Joel, "Charles Steinmetz: Master of Electricity." *Cricket*, May/June 2013.

7. Steinmetz, Charles Proteus, *Elementary Lectures on Electrical Discharges, Waves, Impulses, and Other Transients* (New York: McGraw-Hill, 1911), 26.

8. Hanson, Kip, *Machining for Dummies* (Hoboken, NJ: John Wiley & Sons, 2018).

9. Bertotti, George, *Hysteresis in Magnetism* (San Diego, CA: Academic Press, 1998), 3.

10. Ibid.

11. Asimov, Isaac, *Understanding Physics: Light, Magnetism, and Electricity* (New York: Walker and Company, 1966), 184–185.

12. Gibilisco, Stan, *Beginner's Guide to Reading Schematics* (New York: McGraw-Hill, 2014), 137.

13. Steinmetz, Charles Proteus, *Theory and Calculation of Electric Circuits* (New York: McGraw-Hill, 1917), 113.

14. "On the Law of Hysteresis," *AIEE Trans.*, IX: 3–64, 1892; Proc. of the IEEE, 72 (2): 197–221, doi: 10.1109/PROC.1984.12842.

15. Steinmetz, *Elementary Lectures on Electric Discharges, Waves, Impulses, and Other Transients*, 1–8.
16. Kline, 147.
17. http://eee24h.blogspot.com/2011/04/ac-transient-analysis.html.
18. Kline, 139.
19. Steinmetz, *Theory and Calculation of Electric Circuits.*
20. Kline, 140–141.

Chapter 4: The Current Wars

1. http://library.buffalo.edu/projects/cases/niagara.htm.
2. http://edison.rutgers.edu/topsy.htm.
3. Cawthorne, Nigel, *Tesla: The Life and Times of an Electric Messiah* (New York: Chartwell Books, 2014), 42.
4. Jonnes, Jill, *Empires of Light* (New York: Random House, 2003), 168.
5. Cawthorne, 135–138.
6. Tesla, Nikola, "My Inventions," *Electrical Experimenter*, February 1919.
7. Tesla, Nikola, *Experiments with Alternate Currents of High Potential and High Frequency* (New York: McGraw-Hill, 1904), 4.
8. Jonnes, 163.
9. Cawthorne, 44.
10. Jonnes, 258.
11. Ibid., 294.
12. Ibid., 296–297.
13. http://library.buffalo.edu/projects/cases/niagara.htm.
14. Aldrich, Lisa J., *Nikola Tesla and the Taming of Electricity* (Greensboro, NC: Morgan Reynolds Publishing, 2005).
15. Kline, 19.
16. Lavine, 62–63.
17. Kent, David J., *Edison: The Inventor of the Modern World* (New York: Fall River Press, 2016).
18. http://www.oempanels.com/what-does-single-and-three-phase-power-mean.
19. Steinmetz, *Elementary Lectures on Electrical Discharges, Waves, Impulses, and Other Transients*, 33.
20. Hammond, *Men and Volts*, 231.
21. https://www.platinumelectricians.com.au/blog/importance-grounding-electrical-currents/.

Chapter 5: A Street Lamp Named Steinmetz

1. http://www.fearof.net/fear-of-darkness-phobia-nyctophobia/.
2. http://www.bbc.com/news/magazine-16964783.

3. http://www.edisontechcenter.org/ArcLamps.html.
4. http://www.stouchlighting.com/blog/history-of-street-lighting-in-the-usa.
5. Steinmetz, Charles Proteus, *Radiation, Light, and Illumination* (New York: McGraw-Hill, 1918), 272–273.
6. Kline, 135.
7. Steinmetz, *Radiation, Light, and Illumination*, 158.
8. Lavine, 95.
9. Kline, 306.
10. "The Solar-Powered Road," *Time*, January 23, 2017, 23.
11. Leonard, Jonathan, *Loki: The Life of Charles Proteus Steinmetz* (New York: Doubleday, 1929), 190.
12. https://dailygazette.com/article/2016/05/04/0504_Steinmetz.

Chapter 6: Modern Jove Hurls Lightning in a Lab

1. https://learn.weatherstem.com/modules/learn/lessons/36/02.html.
2. Simon, Seymour, *Lightning* (New York: Scholastic, 1997).
3. https://weather.com/storms/severe/news/united-states-lightning-deaths-2015.
4. http://www.lightningsafety.noaa.gov/odds.shtml.
5. https://www.unbelievable-facts.com/2012/09/lightning-bolt-restores-mans-sight-and.html.
6. https://www.thevintagenews.com/2015/12/09/10-deadly-inventions-in-history/2.
7. Kline, 272.
8. Dray, Philip, *Stealing God's Thunder* (New York: Random House, 2005), 57.
9. Dray, 97–98.
10. Ibid., 91.
11. Steinmetz, Charles Proteus, "The Oxide Film Lightning Arrester," *Proceedings of the American Institute of Electrical Engineers*, June 1918.
12. Maxell, Selby, "Steinmetz First to Produce Artificial Lightning Flashes," *Chicago Daily Tribune*, October 27, 1923.
13. Asimov, Isaac, *Asimov's Biographical Encyclopedia of Science and Technology* (New York: Doubleday, 1964).
14. Kline, 297–298.

Chapter 7: Driven by Electricity

1. https://auto.howstuffworks.com/electric-car.htm.
2. https://www.conserve-energy-future.com/howelectriccarswork.php.
3. https://www.eurosport.com/athletics/how-fast-does-usain-bolt-run-in-mph-km-per-hour-is-he-the-fastest-recorded-human-ever-100m-record_sto5988142/story.shtml.

4. https://energy.gov/articles/history-electric-car.
5. http://cantonasylumforinsaneindians.com/history_blog/tag/cost-of-living-in-1912/.
6. https://jalopnik.com/5564999/the-failed-electric-car-of-henry-ford-and-thomas-edison.
7. Hammond, John, *Charles Proteus Steinmetz: A Biography* (New York: Century, 1924).
8. *The Motley Fool*, email, February 19, 2018.
9. https://www.tesla.com/support/model-s-specifications.
10. https://www.energy.gov/eere/articles/how-does-lithium-ion-battery-work.
11. http://mashable.com/2013/01/17/tesla-electric-car/#IQ0k4kM3MZqT.
12. https://www.washingtonpost.com/news/the-switch/wp/2014/12/30/heres-how-far-every-tesla-model-s-can-go-on-a-single-charge/.
13. Solomon, Dan, "The Future's So Bright," *Smithsonian*, April 2018, 70.
14. Palm Beach Research Group, June 1, 2018.

Chapter 8: An Exaltation of Geniuses

1. Natarajan, Priyamrada, "Calculating Women," *New York Review of Books*, May 25, 2017, 38.
2. https://www.tinypulse.com/blog/sk-company-culture-of-collaboration.
3. *Solutions for Private Equity Firms* (PowerPoint). (College Station, TX: Flippen).
4. http://www.gettyimages.com/detail/news-photo/schenectady-new-york-charles-proteus-steinmetz-famous-news-photo/515996572#61922schenectady-new-yorkcharles-proteus-steinmetz-famous-american-picture-id515996572.
5. Golder, 26.
6. Cawthorne, 53.
7. Lavine, 222.
8. http://www.gettyimages.com/detail/news-photo/schenectady-new-york-charles-proteus-steinmetz-famous-news-photo/515996572#61922schenectady-new-yorkcharles-proteus-steinmetz-famous-american-picture-id515996572.
9. http://www.who2.com/bio/leonardo-da-vinci/.
10. https://www.linkedin.com/pulse/relatively-speaking-when-albert-einstein-came-ge-tomas-kellner/.
11. https://www.ge.com/reports/post/87725858315/einsteins-relativity-will-make-your-electricity/.
12. http://www.elp.com/articles/powergrid_international/print/volume-21/issue-1/features/the-internet-of-things-connection-to-the.html.
13. *SmartGrid Today*, November 11, 2017.
14. https://spectrum.ieee.org/tech-talk/telecom/internet/popular-internet-of-things-forecast-of-50-billion-devices-by-2020-is-outdated.

15. Steinmetz, *Four Lectures on Relativity and Space*, v.

16. https://www.business2community.com/us-news/thomas-edisons-mother-lying-contents-expulsion-letter-largely-false-01719324#hpPMcBRFIfoD23rA.97.

17. Lavine, 5–6.

18. Kent, 177–180.

19. Ibid., 129–134.

20. *The Writer's Life*, December 15, 2017.

21. Kent, 245.

22. Jonnes, 113.

23. Kline, 246.

24. Piosczyk, Hannah, et al, "Prolonged Sleep under Stone Age Conditions," *Journal of Clinical Sleep Medicine* 10, no. 7 (2014).

25. https://www.health.harvard.edu/press_releases/light-from-laptops-tvs-electronics-and-energy-efficient-lightbulbs-may-harm-health.

26. Cawthorne, 138.

27. Golder, 42.

28. Bleiler, E. F., *Science Fiction Writers* (New York: Scribner, 1982), 41.

Chapter 9: The Bohemian Scientist

1. Leonard, 239.

2. http://gremsdoolittlelibrary.blogspot.com/2013/04/happy-birthday-charles-steinmetz.html.

3. Lavine, 220.

4. https://www.union.edu/about/.

5. https://psyarxiv.com/edzda.

6. https://en.wikiquote.org/wiki/Joseph_Priestley.

7. http://www.telegraph.co.uk/news/religion/9102740/Richard-Dawkins-I-cant-be-sure-God-does-not-exist.html.

8. https://en.wikiquote.org/wiki/Richard_Dawkins.

9. https://en.wikiquote.org/wiki/Stephen_Hawking#God_Created_the_Integers_(2007).

10. Hammond, *Steinmetz: A Biography*, 445–448.

11. Hart, 17.

12. http://www.foxnews.com/tech/2011/11/17/radio-days-are-back-ham-radio-licenses-at-all-time-high.html.

13. http://www.newsfinder.org/site/more/radio_broadcast/.

14. http://www.vvara.org/clubinfo/pdf/History_of_Ham_Radio.pdf.

15. https://www.misci.org/archives/finding_aids/Steinmetz.htm.

16. Froehlich, Fritz E. and Allen Kent, *The Froehlich/Kent Encyclopedia of Telecommunications Volume 15* (New York: Marcel Dekker, 1998), 467.

17. https://www.smithsonianmag.com/smart-news/inventor-telegraph-was-also-americas-first-photographer-180961683/.
18. https://www.sciencelearn.org.nz/resources/960-the-nitrogen-cycle.
19. "Charles P. Steinmetz: Electrical Expert," *Public Service Magazine*, December, 1909.
20. Steinmetz, *Radiation, Light, and Illumination.*
21. Miller, 12.
22. Ibid., 51.
23. Ibid., 67–69.
24. https://www.omniglot.com/writing/shorthand.htm.
25. Leonard, 234–237.

Chapter 10: Fun, Games, and Over

1. Leonard, 274.
2. http://www.edisontechcenter.org/MercuryVaporLamps.html.
3. Leonard, 220–222.
4. Lavine, 178–179, 200, 202.
5. Ibid., 197.
6. Ibid., 235–236.
7. Kline, 291.

Glossary

Alternating current (AC): Electric current that repeatedly changes direction of flow many times per minute.

Alternator: A generator producing alternating current.

Anode: Positively charged electrode of a battery.

Arc: An electric arc is a discharge between two electrodes within a sealed glass container. It is produced by electrical current ionizing gas within the container. The discharge produces illumination.

Armature: A mechanism consisting of copper coils wound on an iron drum that turns between the poles of a magnet.

Capacitance: Measured in coulombs per volt, capacitance is the quantity of electrical charge divided by the electrical potential (voltage difference).

Capacitor: Two metal plates in proximity, with one positively charged and the other negatively charged; also called a condenser.

Cathode: Negatively charged electrode.

Conductor: Any material that readily carries an electric current.

Direct current (DC): Electric current that continuously flows in one direction only.

Dynamo: A generator that uses a rotary electrical switch to produce direct current.

Electric current: The flow of electrons through a conductive material such as copper, or more generally the movement of any charged particle through any conductive medium, such as ions through an electrolyte.

Electric potential difference: The difference in electric potential energy between two charged bodies.

Electrolyte: Fluids that create current through movement of ions through the fluid from one electrode to its opposite.

Electromagnetism: The fundamental force encompassing two related phenomena, electricity and magnetism.

Generator: A machine that produces electrical power.

Ground wire: A wire connected to the ground so that if insulation is worn and wires are exposed, touching the bare wire does not produce a dangerous shock.

Impedance: Resistance of an AC circuit to the flow of current.

Induction: Creating electricity by rotating a copper coil through a magnetic field.

Insulator: Rubber, glass, and any other material with substantial electrical resistance that blocks the flow of current; wires and other conductors are wrapped in insulators to prevent humans and equipment from being harmed by static shock.

Ion: A particle with either a positive or negative electrical charge.

Lightning: Discharge of a bolt of electricity from a cloud either to another cloud or to the ground or objects and structures on the ground.

Load factor: The amount of electricity carried by a circuit, transmission line, or grid.

Magnet: A material in which the electrons are polarized or aligned, causing the material to attract certain metals.

Polyphase current: Current with multiple wavelengths, frequencies, or phases.

Potentiometer: A device for measuring voltage against a set point or threshold value.

Renewable energy: An energy sourced that is for all practical purposes inexhaustible and can never run out; examples include solar energy, wind, geothermal, and hydroelectric.

Resistance: The characteristic of a material that blocks the flow of a direct current.

Static electricity: An electric charge, typically produced by friction, that causes sparks as it suddenly jumps from point A to point B.

Transformer: An electrical apparatus that either "steps up" (increases) or "steps down" voltage in electric current. Stepping up enables longer distance transmission, and stepping down ensures the voltage level does not damage the devices powered by it.

Turbine: A machine in which a blade, usually driven by steam pressure, spins an armature or shaft, converting the heat energy of the steam into the kinetic energy of movement and, as the armature moves within a magnetic field, further converts the kinetic energy into electrical energy.

Volt: Unit of measure of the electromotive force driving an electric current through a wire or other conductor.

Resources

IEEE Spectrum
The monthly trade journal published by the Institute for Electrical and Electronic Engineers
https://spectrum.ieee.org/

Power Markets Today
An e-newsletter for the utility industry
www.powermarketsday.com

Smart Grid Today
A journal for the utility industry
www.smartgridtoday.com

Bibliography

Aldrich, Lisa J. *Nikola Tesla and the Taming of Electricity* (Greensboro, NC: Morgan Reynolds Publishing, 2005).

Asimov, Isaac. *Asimov's Biographical Encyclopedia of Science and Technology* (New York: Doubleday, 1964).

Asimov, Isaac. *Asimov's Chronology of the World* (New York: HarperCollins, 1991).

Asimov, Isaac. *Understanding Physics* (New York: Walker and Company, 1966).

Broderick, John Thomas. *Steinmetz and His Discoveries* (Schenectady, NY: Robson & Adee, 1924).

Caldecott, Ernest and Philip Langdon Alger, eds. *Steinmetz the Philosopher* (Mohawk Development Service, 1965).

Cawthorne, Nigel. *Tesla: The Life and Times of an Electric Messiah* (New York: Chartwell Books, 2014).

Drat, Philip. *Stealing God's Thunder* (New York: Random House, 2005).

Franklin, Benjamin. *The Autobiography of Benjamin Franklin* (Millennium, 2015).

Furgang, Kathy. *Zoom in on Power Grids* (New York: Enslow, 2018).

Garlin, Sender. "Charles Steinmetz: Scientist and Socialist." In *Three American Radicals: John Swinton, Charles P. Steinmetz, and William Dean Howells* (Boulder, CO: Westview Press, 1976).

Gibilisco, Stan. *Beginner's Guide to Reading Schematics* (New York: McGraw-Hill, 2014).

Gilbert, John. "Collectivism and Charles Steinmetz," *Business History Review* 48, no. 4 (Winter 1974): 520–520.

Hammond, John. *Charles Proteus Steinmetz: A Biography* (New York: Century, 1924.

Hammond, John. *Men and Volts: The Story of General Electric* (Philadelphia: J. B. Lippincott, 1941).

Hart, Larry. *Schenectady's Golden Era: 1880–1930* (Schenectady, NY: Old Dorp Books, 1974).

Hynes, Patricia. *Power Up: Electric Power Grid* (North Mankato, MN: Cherry Lake Publishing, 2008).

Jonnes, Jill. *Empires of Light: Edison, Tesla, Westinghouse, and the Race to Electrify the World* (New York: Random House, 2003).

Kent, David J. *Edison: The Inventor of the Modern World* (New York, Fall River Press, 2016).

Kline, Ronald. *Steinmetz: Engineer and Socialist* (Baltimore: Johns Hopkins University Press, 1992).

Lavine, Sigmund A. *Steinmetz: Maker of Lightning* (New York; Dodd, Mead, 1955).

Leedskalnin, Edward. *Magnetic Current* (Eastford, CT: Martino Fine Books, 2011).

Leonard, Jonathan. *Loki: The Life of Charles Proteus Steinmetz* (New York: Doubleday, 1929).

Lowe, Derek. *The Chemistry Book* (New York: Sterling, 2016).

Martin, Joel and William Birnes. *Edison vs. Tesla: The Battle over Their Last Invention* (New York: Skyhorse, 2017).

Maxwell, Selby. "Steinmetz First to Produce Artificial Lightning Flashes," *Chicago Daily Tribune*, October 27, 1923.

Miller, John. *Modern Jupiter: The Story of Charles Proteus Steinmetz* (New York: American Society of Mechanical Engineers, 1958).

Nagpurwala, Q. H. "Design of Steam Turbines. (Lecture, M. S. Ramaiah School of Advance Studies, Bangalore).

Remscheid, Emil J. and Virginia Remscheid Charves. "Recollections of Steinmetz: A Visit to the Workshops of Dr. Charles Proteus Steinmetz" (General Electric Company, Research and Development, 1977).

Schuh, Mari. *Magnetism* (New York: Scholastic, 2011).

Simon, Seymour. *Lightning* (New York: Scholastic, 1997).

Solomon, Robert. *The Little Book of Mathematical Principles* (Grantham, Lincolnshire: IMM Lifestyle Books, 2016).

Steinmetz, Charles Proteus. *Elementary Lectures on Electric Discharges, Waves, Impulses, and Other Transients* (New York: McGraw-Hill, 1911).

———. *Theory and Calculation of Electric Circuits* (New York: McGraw-Hill, 1917)

Tesla, Nikola. "My Inventions," *Electrical Experimenter*, February 1919.

The Utility Business Customer Survey on Energy Management (Spokane, WA: Ecova, 2017).

Illustration Credits

Fig. 1-1. Dave McCoy.

Fig 1-2. Public domain.

Fig. 1-3. *The World's Work, Volume XLIV* (New York: Doubleday, 1922).

Fig. 1-4. German Federal Archives, 1890.

Fig. 1-5. Public domain.

Fig. 1-6. Steinmetz, Charles Proteus. *Theory and Calculation of Electric Circuits* (New York: McGraw-Hill, 1917).

Fig. 1-7. Ibid.

Fig 2-1. Dave McCoy.

Fig. 2-2. Public domain.

Fig. 2-3. Robert W. Bly.

Fig. 2-4. Dave McCoy.

Fig. 2-5. Ibid.

Fig. 2-6. Ibid.

Fig. 2-7. Public domain.

Fig. 2-8. Ibid.

Fig. 2-9. Dave McCoy.

Fig. 2-10. Ibid.

Fig. 2-11. Ibid.

Fig. 2-12. Ibid.

Fig. 2-13. Steinmetz, *Theory and Calculation of Electric Circuits.*

Fig. 3-1. Dave McCoy.

Fig. 3-2. Steinmetz, Charles Proteus. *Elementary Lectures on Electrical Discharges, Waves, Impulses, and Other Transients* (New York: McGraw-Hill, 1911).

Fig. 3-3. Gibilisco, Stan. *Beginner's Guide to Reading Schematics* (New York: McGraw-Hill, 2014).

Fig. 3-4. Steinmetz, *Elementary Lectures on Electrical Discharges.*

Fig. 3-5. Dave McCoy.

Fig. 3-6. Steinmetz, *Theory and Calculation of Electric Circuits.*

Fig. 3-7. Dave McCoy.

Fig. 4-1. Public domain.

Fig. 4-2. Ibid.

Fig. 4-3. Joseph G. Gessford.

Fig. 4-4. Public Domain.

Fig. 4-5. Ibid.

Fig. 4-6. Ibid.

Fig. 4-7. Ibid.

Fig. 4-8. Courtesy iStock.

Fig. 5-1. U.S. Patent Office.

Fig. 5-2. Dave McCoy.

Fig. 5-3. Public domain.

Fig. 5-4. Steinmetz, *Theory and Calculation of Electric Circuits.*

Fig. 5-5. Schenectady County Historical Society.

Fig. 5-6. Public domain.

Fig. 6-1. Dave McCoy.

Fig. 6-2. Ibid.

Fig. 7-1. Courtesy Tesla, Inc.

Fig. 7-2. Dave McCoy.

Fig. 8-1. Ibid.

Fig. 8-2. Public domain.

Fig. 8-3. Courtesy iStock.

Fig. 8-4. Library of Congress.

Fig. 8-5. Harry Shipler.

Fig. 8-6. Ferdinand Schmutzer.

Fig. 8-7. Louis Bachrach.

Fig. 8-8. Dave McCoy.

Fig. 9-1. Courtesy Union College. Reprinted with permission.

Fig. 9-2. Dave McCoy.

Fig. 9-3. Ibid.

Fig. 9-4. Public domain.

Fig. A-1. Dave McCoy.

Index

About the Author

Bob Bly has nearly four decades of experience as a science and technology writer. He is the author of more than 95 books including *The Science in Science Fiction* (BenBella) and *A Dictionary of Computer Words* (Banbury).

Bly holds a BS in chemical engineering from the University of Rochester and is a member of the American Institute of Chemical Engineers. He is also trained as a Certified Novell Administrator. Before becoming a full-time freelance writer, Bly was a technical writer for Westinghouse and advertising manager for Koch Engineering, an industrial manufacturer.

Bly has written articles, newsletters, and other technical materials on electricity and electronics for such organizations as GE Semiconductor, RCA, BOC Gases, Schneider Electrical, IEEE, and Leviton. His articles have appeared in *Science Books & Films*, *City Paper*, *New Jersey Monthly*, *Chemical Engineering Progress*, and many other magazines and newspapers. His science web site is **www.mychemset.com** and his writing web site is **www.bly.com**.